POWER AND GENDER IN EUROPEAN
RURAL DEVELOPMENT

Perspectives on Rural Policy and Planning

Series Editors:
Andrew Gilg
University of Exeter, UK
Keith Hoggart
King's College, London, UK
Henry Buller
Cheltenham College of Higher Education, UK
Owen Furuseth
University of North Carolina, USA
Mark Lapping
University of South Maine, USA

Other titles in the series

Power and Gender in European Rural Development

Edited by

HENRI GOVERDE
University of Nijmegen and Wageningen University, The Netherlands
HENK DE HAAN
Wageningen University, The Netherlands
MIREIA BAYLINA
Autonomous University of Barcelona, Spain

Routledge
Taylor & Francis Group

LONDON AND NEW YORK

First published 2004 by Ashgate Publishing

Reissued 2018 by Routledge
2 Park Square, Milton Park, Abingdon, Oxon OX14 4RN
711 Third Avenue, New York, NY 10017, USA

Routledge is an imprint of the Taylor & Francis Group, an informa business

First issued in paperback 2018

A Library of Congress record exists under LC control number: 2003058366

Notice:
Product or corporate names may be trademarks or registered trademarks, and are used only for identification and explanation without intent to infringe.

Publisher's Note
The publisher has gone to great lengths to ensure the quality of this reprint but points out that some imperfections in the original copies may be apparent.

Disclaimer
The publisher has made every effort to trace copyright holders and welcomes correspondence from those they have been unable to contact.

ISBN 13: 978-0-815-39112-8 (hbk)
ISBN 13: 978-1-138-62020-9 (pbk)
ISBN 13: 978-1-351-15148-1 (ebk)

Contents

List of Contributors

Mireia Baylina is an Associate Professor in the Department of Geography, Autonomous University of Barcelona, Spain.

Nina Gunnerud Berg is an Associate Professor in the Department of Geography at the Norwegian University of Science and Technology (NTNU) in Trondheim, Norway.

Bettina Bock is an Associate Professor in the Rural Sociology Group, Department of Social Sciences at Wageningen University, the Netherlands.

Alia Gana is a Ph.D. in Rural Sociology from Cornell University and currently Associate Professor at the College of Agriculture, University of 7 November Carthage, Tunisia.

Maria Dolors Garcia-Ramon is Professor at the Department of Geography, Autonomous University of Barcelona, Spain.

Henri Goverde is Associate Professor in Public Administration in the Faculty of Management Sciences at the University of Nijmegen, and Professor in Political Science, Department of Governance and Law at Wageningen University, the Netherlands.

Henk de Haan is Associate Professor in the Department of Social Spatial Analysis at Wageningen University, the Netherlands.

Torsti Hyyryläinen is a Senior Researcher at the Mikkeli Institute for Rural Research and Training at the University of Helsinki in Finland.

Lutz Laschewski is a Lecturer for Agricultural Policy and Food Marketing at the Institute for Agricultural Economics and Process Engineering, Faculty of Agricultural and Environmental Sciences of the University of Rostock, Germany.

Hans Kjetil Lysgård is a Senior Researcher at the Agder Research Foundation in Kristiansand, Norway.

José Ramón Mauleón is a Lecturer in the Department of Sociology 2, University of the Basque Country in Bilbao, Spain.

Rosemarie Siebert is a Researcher at the Centre for Agricultural Landscape and Land Use Research in Müncheberg, Germany.

Rachel Woodward is a Lecturer in the School of Agriculture, Food and Rural Development at the University of Newcastle upon Tyne, United Kingdom.

Acknowledgements

This book is the result of a series of conferences and seminars organized by the participants of the Action Group 'Leadership, local power, and restructuring rural communities' (AG 5). This group consisted of members from eleven European countries, and was one of the five groups working within the European COST (European Cooperation in the field of Scientific and Technical Research) Action A12 on Rural innovation. The contributors to this volume acknowledge the financial support from the COST programme, and thank Imre Kovach for his role as a coordinator and stimulator of our group. The following institutions kindly provided the facilities for our three meetings: the Mikkeli Institute for Rural Research and Training at the University of Helsinki; UMR LADYSS CNRS 'Laboratoire des Dynamiques Sociales et Recomposition des Espaces' University of Paris X, and the Department for Political Science at the Hungarian Academy of Sciences in Budapest. We especially thank Tuula Isouo, Päivi Pylkanen, Torsti Hyyrylainen, Marie-France Epagneul, Nicole Matthieu, and Franciska Veber for their collaboration and hospitality. In preparing the manuscript we received invaluable support from Michelle Mellion, Catharina de Kat-Reynen and Sylvia Belmonte.

Henri Goverde
Henk de Haan
Mireia Baylina

INTRODUCTION

Chapter 1

Introduction: Power and Gender in European Rural Development

Henri Goverde, Mireia Baylina, Henk de Haan

Within the global framework of rural innovation and rural development, this book analyses general European trends in rurality in the face of power and gender. We start from the assumption that besides the traditional key actors in rural areas (farmers and their interest groups as well as politicians and policy-makers), new ones have entered the political, social, economic and cultural arenas. Together these actors have produced a new rural context, which has become extremely complex. In accordance with the Cork Declaration (1996) we also assume that a new contract between agriculture and society will not only include agreements on traditional products (milk, cereals, beef) and production factors (land, labour, capital, knowledge), but also consumption items such as landscape, nature, environment, rural development, food safety and, at a more abstract level, space, biosystems, production methods and conditions (regarding animal welfare and genetic modification), ethics and morality (Vihinen, 2001).

Viewed from the perspective of this new contract, the (policy-oriented) research question arises whether or not rurality and rural development will face fundamental new problems that demand urgent rural innovation. In the words of the Budapest Declaration on Rural Innovation (2002, p. 8): 'To become a useful conceptual instrument in research, "innovation" may be analysed with regard to its effects ... to (re-emphasize to those concerned with rural development that social (technical, economic) change is not neutral, in either its origins or its outcomes, to inequalities of gender and power, at local or other levels of social life.' The related academic question is how these changes affect the lives of rural people and the structure and functioning of rural societies. This book focuses on one of the most relevant aspects of (rural) society: the dynamics in (local) power relations. While Part I uses a wide scope to present types of local power relations in different rural contexts, Part II focuses on rural gender relations, particularly by reflexive analysis of (academic) studies in their dynamic socio-economic and political contexts.

Rurality and Discourse

What is happening in rural areas can only be understood theoretically as the combined effect of local and global interactions ('glocalization') and the role of spatially dispersed agency. In rural studies, 'the rural' is no longer taken for granted as an objective fact, but rather as a plurality of social and political constructions and /or representations of spatial qualities. It is significant that while academic discourse is beginning to reject the usefulness of 'rural' as a scientific concept, it is going through a phase of revival at all levels of political and popular discourse. These discourses range from abstract philosophies about the great potentials of rural amenities for society at large, to reference by tourists to attractive places in the countryside. For policy-makers, rural development agencies, investment companies and the general population, the rural constitutes a reality as a natural and cultural *décor* for the performance of specific activities and for experiencing the rural sensation.

Rural discourses are not, however, without practical significance. Policy discourses, the production of images by the media and other popular representations are somehow reflected in concrete political measures, planning devices, migration movements, travelling destinations, investment decisions and so on. These are all based on conceptions of how rural space should be ordered and how to arrange urban–rural relations in order to meet predefined individual and social demands. Rural discourses thus reflect a plurality of interests, lifestyles and identities, which translate into real *in situ* spatial transformations.

However, rurality is not only a cultural theme of reflection and appropriation of otherness. The concept is not restricted to talking and thinking 'about rurality' as a resource to be mobilized for specific interests. The object of discourse is at the same time the living space for people who are mostly not actively preoccupied with the significance of their living space for a wider audience and market. They are, however, affected by the externally induced transformations in terms of new opportunities, restrictions and the impact of newcomers, visitors and so on. These influences, together with living–space-related experiences result in localized rural discourses or discourses '*in* rurality.' Perhaps the difference between discourse 'about rurality' and 'in rurality' is that the former takes an urban perspective on rurality, while the latter has a rural perspective on urbanity.

Rurality: Local Power and Gender

Attention has also been paid to the impacts of discourses 'about' and 'in' rurality, particularly to how the interactions between established interests and newcomers affect the local power relations. These relationships will be conceptualized below. Of course, the concept of power itself has always been contested. That is why this concept will also be further elaborated below.

Gender is supposed to be a social structure, like class and nationality, which ensures that political influence within a society is unequally divided (Faulks, 1999, pp. 1, 118, 124). Therefore, gender relations are basic power relations in urban as well as in rural societies. For that reason, we introduce a gender perspective on rural (local) power relations in order to see how the overall changing structure of power relations affects the position of men and women, particularly at the local and regional level of governance. Certainly, the emerging interest in new rural governance provides an opportunity to start examining the characteristics of women's and men's contribution to the policy process at the local level. Moreover, the gendered nature of rural decision-making itself will be investigated. From this perspective, relevant research questions are 'which biases are mobilized,' 'which mechanisms produce the structural underexposure of the difference between men's and women's interests,' and 'what are the impacts of these phenomena on the power relations in rural areas.'

The chapters collected in this book are written by academics who are interested in the impacts of the interactions between established interests and newcomers in local power relations. The book has its origin in the EU Cost A12 programme 'Rural Innovation' (1998–2002) – Working Group 5 'Leadership, Local Power, and Rural Restructuring.' The setting of this EU Cost programme offered the opportunity to set up a network of scientists, which resulted in the book *Leadership and Local Power in European Rural Development* (Halfacree, Kovach and Woodward, 2002). *Rurality, power and gender* is the second outcome of this network. The book is structured in two parts. The first part, 'Power and rural development in action', consists of five chapters. The second part, 'Gender and rurality in academic discourse', also has five chapters, which are all devoted to the analysis of the scientific production of rural studies from a gender perspective in different European and Southern Mediterranean countries. The chapters in Part I are based on a mixture of data collection and analysis methods. Extensive use is made of all kinds of government reports and other policy-relevant documents. Moreover, geographical and sociological survey research techniques are used in several chapters. Almost all contributions include interviews with key actors. Part II is mainly based on literature analysis, mostly using academic and also non-academic reports on rural gender issues. It should be noted, however, that all authors in Part II also have extensive experience themselves with empirical research on gender and rural development.

Part I: Power and Rural Development in Action

The main argument of the chapters in Part I is that traditional and new forms of social organization in rural areas can be perceived as elements or constructions based on social capital, which is continually changing and thus creates new forms of (political) participation in rural society. Sometimes and in some places these new forms of participation seem to be rather gendered in character. Furthermore, the innovative character of these new forms of rural organization can be questioned. Perhaps there are only changes in rhetoric and signs of symbolic policy-making. Changing forms of social capital represent the dynamics of (local) power, which influences not only the relation between state and civil society but also gender relationships in rural areas. The dynamics of power are also studied in the context of changing social formations, while some contributions approach power dynamics from the perspective of power/knowledge systems, which produce competing discourses about rurality. The competing character of discourses can obstruct the congruency between (sectoral) policies in multi-level governance concerning rural development (see Figure 1.1).

Figure 1.1 The dynamics of power and rural development

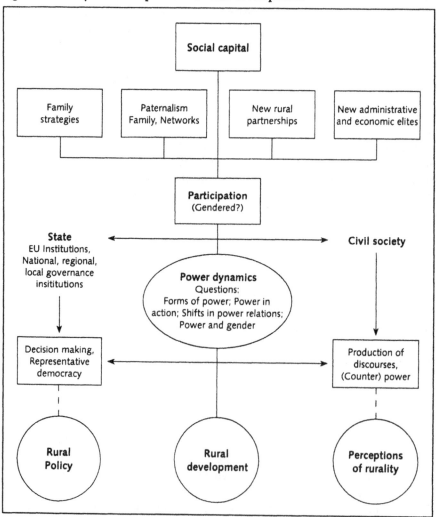

A critical remark concerning this scheme and its line of argument is the hidden assumption that social capital and political capital are synonymous. Furthermore, it seems to assume that the relationship between social trust and social capital, on the one hand, and political trust and political capital, on the other, is simple or straightforward. It is important to realize, however, that such a relationship exists, as social capital and civil society theory predicts, but not at the individual level, and only in a complicated and indirect manner at the system level. As Newton (2001, pp. 211–212) has explained '... since social capital and civil society are essentially a social and collective property of social systems, not a characteristic feature that individuals carry around with them, the relationship is found at the aggregate level of society as a whole. Consequently, while little can be predicted about an individual's political trust or confidence in public institutions on the basis of his or her social trust, countries with relatively high levels of social trust tend to have relatively high levels of political confidence as well.' By and large the reason seems to be that the relationship between individual social trust and political trust is mediated by the effectiveness of social and political institutions. 'Individual social trust helps to build the cooperative social relations on which effective social and political organizations are built – *a bottom-up process*. Effective social and political organizations help to create effective and legitimate government, which then help to create the social conditions for high levels of social capital and a well-developed civil society – *a top-down process*. Similarly, individual political trust helps to build effective political institutions that enable governments to perform well, and to build up political capital as well as creating the conditions for a flourishing civil society – another bottom-up and top-down process. Hence, social capital and a developed civil society help to make good government possible, and good government helps to sustain social capital and the conditions of civil society' (Newton, 2001, p. 211).

Rural development, and particularly rural innovation, is not a neutral, non-political process. On the contrary, rural innovation will be beneficial to some rural groups, whereas others will experience disadvantages. Those who are included in the political process have more opportunities to satisfy their interests in comparison to all the minorities that are politically marginalized or completely excluded. Viewed from this perspective, a major issue of power is related to the differences between men and women in access to the political, economic and social arena. In many countries the access to and the use of power resources is gendered at the national level of governance, as well as in regional and local processes of decision-making (Bock, 2002). Of course, when the relative exclusion from political and policy processes has been institutionalized, it prohibits the empowerment of certain groups and as such it is an obstacle to participation in bottom-up processes in (rural) innovation.

On the other hand, effective rural innovation should not be theorized from the idea of the uniformity of rural development or its purely economic character. Such an approach has been the core of the modernization project since the end of World War II. Existing EU rural development policies have tended to be dominated by a singular perspective and discourse about what constitutes rurality. Rural innovation should take into account the potential sustainability of differences in European rural regions. From this perceptive, rural development is above all a heuristic device (Van der Ploeg et al., 2000). Therefore, an analysis of rural development practices across a wide range of countries and regions can enhance insights into the variety of socially and culturally innovative development practices and experiments at the regional and local governance level. In this way the relevance is shown of shifts in the distribution of power resources, the appearance of new actors in rural areas, new alliances between actors, empowerment of marginalized individuals and groups, and the mutual dependency and support in regional/local business chains. The concepts of power will be further described and analysed as described in the introduction to Part I.

Part II: Gender and Rurality in Academic Discourse

According to the authors in Part II, economic and political transformations in rural areas are shaped by the male and female relationships in the household and the community, and in turn, the implications of these changes are considerably different for men and women. As mentioned above, the chapters in part II consider the way in which the gendered dimension of rural development is reflected in academic discourses in de period after World War II. Through these academic discourses one can deduce the different phases of the involvement of women in rural development as well as the multiple dimensions of gendered power. This part of the book includes five chapters from different countries: the United Kingdom, the Netherlands, Norway, Tunisia, and Spain. Although these reviews cover a substantial amount of academic literature, the authors do not claim exhaustiveness.

Over the past thirty years (mostly female) scientists have expanded the legitimate territory of inquiry in many disciplines (sociology, geography, anthropology, political science) to the study of feminist and women's issues. The intention of the chapters in Part II is to describe socio-economic and cultural trends and issues in the relationships between gender and rural restructuring, and to document the different conceptual and theoretical approaches used in particular projects. In general, these studies claim that the theoretical endeavours provide an informed and coherent body of knowledge on rural areas, from which it is possible to make proposals for development and innovation. These chapters illustrate the cultural diversity of rural Europe, and suggest that theories and policies of rural restructuring have to consider the particularity of social contexts and places. From the results, we can see that a relatively rich literature on gender and rural development has been developed in almost all European countries as well as in some countries of the southern Mediterranean area. The literature shows also a sequential move in theoretical arguments. The first phase includes a very political project to make women economically visible. In the next phase there is a focus on explaining differences and inequalities oriented towards empowering women as social agents. Finally, there is the birth of the notion of 'gender identity,' claiming a 'plurality' in gender and development. These foci of attention sometimes co-exist, while some of them may not have been relevant at all. Nonetheless, European countries show a remarkable coherence regarding the way in which academic discourses developed, and how different gender-related problems were successively 'discovered' and researched. There are, however, also remarkable differences, showing the impact of specific academic traditions and the socio-cultural differences between regions and countries.

How can we understand rural development and rurality nowadays, particularly from a power and gender perspective? The chapters in this book bring us much closer to possible answers to this question. It must be said, however, that most of its authors have experienced in the last decade a lack of sufficient resources for relevant longitudinal analysis. It has therefore not been an easy task to deduce trends from regular and continuous data inputs about real practices and cultures in the rural areas. Furthermore, some countries do not produce data aggregated at the scale of local communities and regional institutions at all. In general, throughout Europe there is a lack of systematic data collection about developments in social cohesion, social exclusion, local leadership, social movements, and local power in rural areas. However, some chapters suggest that the Cork Declaration did not announce fundamental new social and political trends, but rather it marked the beginning of a new era in policymaking concerning the rural areas in Europe (see Budapest Declaration). Rural innovation can no longer be perceived as a matter of techniques and instrumental mechanisms only. Real rural innovation should bring 'politics' back into the discussion. That implies a demand for a more critical and radical emphasis on power in general and power in a gendered perspective in particular.

References

Bock, B. (2002), *Tegelijkertijd en tussendoor. Gender, plattelandsontwikkeling en interactief beleid.* Wageningen: Wageningen University (Ph.D thesis).

Faulks, K. (1999), *Political Sociology. A Critical Introduction.* Edinburgh, Edinburg University Press.

Halfacree, K., I. Kovach and R. Woodward (eds) (2002), *Leadership and Local Power in Contemporary Rural Europe.* Aldershot, Ashgate.

Mormont, M. et al., (2002), *The Budapest Declaration on Rural Innovation*, Final Report of EU Cost A12 Action 'Rural Innovation.' Budapest/Brussel/Arlon.

Newton, Kenneth (2001), Trust, social capital, civil society, and democracy. *International Political Science Review*, 22 (2) pp. 201–214.

Ploeg, J.D. van der et al. (2000), Rural development: from practices and policies towards theory. *Sociologica Ruralis* 40 (4) pp. 391–408.

Vihinen, Hilkka (2001), *Recognising Choice. A Study of the Changing Politics and the Common Agricultural Policy trough an Analysis of the MacSharry Reform Debate in Ireland and the Netherlands.* Helsinki, Agrofood Research Finland, Economic Research (MTTL).

PART I

POWER AND RURAL DEVELOPMENT IN ACTION

Chapter 2

Introduction to Part I:
Power and Rural Development in Action

Henri Goverde

Rural development, and particularly rural innovation, is not a neutral, non-political process. On the contrary, rural development can be perceived as the product of the power dynamics in rural areas. In Part I it is assumed that different forms of social capital result in different types and forms of political participation. Social capital, like any form of capital, is unevenly distributed in civil society. The opportunities to accumulate social capital depend not only on agency skills of an individual actor, but they are also embedded in social structures of class, race, ethnicity and gender (Faulks, 1999, pp. 1, 110). For example, it may be expected that men's social capital is of the type that grants access to more respected and (thus) more effective types of political participation. Research has proven time and again that women and men not only differ in degree of political participation but even more in the kind of political behaviour and type of political arena they are engaged in. Men are more active in 'big politics', i.e., in formal, official political fora, whereas women seem to be more often involved in 'small politics', i.e., in informal, unofficial and rather invisible political fora (Siim, 2000; Lovenduski and Stephenson, 1999). There are indications that this is plausible, especially in rural politics (Little, 2002; Shortall, 2002). New rural partnerships are – in general – not rooted in family life nowadays. However, in some European rural areas the relation between family organization, political participation and success or failure of governmental policy implementation is still significant. In the meantime different types of alliances between the established residents and the newcomers in rural areas can often be fruitfully perceived according to gender characteristics.

Participation in rural affairs develops in the interchange between state and civil society relationships (Faulks, 1999). Of course, the 'market' is also relevant as far as it conditions the rules of the exchange of power resources as well as the attributes of the dominant political-economic system. However, these market conditions are treated here mostly as a context.

On the one hand, the dynamics of power at local, national, transnational/EU state levels constitute the institutional framework within which everyday practices of rural development and rural innovation take place. All contributions in Part I take into consideration state institutions of representative democracies and administration of justice in EU member states, which produce relatively comparable rural policies. On the other hand, the dynamics of power themselves can be deduced from different forms of social capital and political participation in specific rural areas. By doing so, shifts in actual power relations can be revealed and analysed. Although many research methods are represented in the contributions of Part I, it is discourse analysis which is particularly fruitful to describe and explain the production of story lines and narratives, which often function as a (counter) power to the policies of the political administrative system. Competing perceptions of rurality are relevant for explaining success and failures of rural policies in practice. State-centered rural policies, on the one hand, and perceptions of rurality in everyday rural practice, on the other hand, form the bandwidth within which power dynamics produce the actual process of rural development. However, the concept of 'power dynamics' is a complex one that needs to be elaborated here.

Concept of Power Dynamics

For the purpose of this book it is assumed that the concept of power is fruitful to explore the political dimension of rural development and gender issues in rural innovation. According to Goverde and van Tatenhove (2000, pp. 106–108) the concept of power consists of three interconnected layers:

- Power as a capacity.
- Power as a relational phenomenon.
- Power as a structural, even systemic phenomenon.

'Power as a capacity' is the most apparent and visible type of power. It concerns 'power over' and consists of the subordination of one person's will by the will of another. Power perceived in this way, called 'episodical power' (Clegg, 1989, pp. 211–218) or 'transitive power' by Goehler (2000, p. 43), restricts the possibilities of actor B (the subaltern) by the will of actor A (the principal). This layer of power is linked to the idea of the production of causal effects. This idea of power as causal power is also integral to the very idea of human agency (Scott, 2001, p. 1). To be an agent is to exercise causal powers that produce specific effects in the world: 'transformative capacity' possessed by human agents (Giddens, 1982). Power as a capacity of social agents, agencies and institutions refers to the way in which the social and physical (rural) environment is maintained and transformed, '... and it concerns the resources which underpin this capacity and the forces that shape and influence its exercise' (Held, 1995, p. 170). Thus, at this first layer of power, the emphasis is on 'power resources' and how agents use them directly, either in asymmetrical or in symmetrical relationships.

According to Scott (2001, pp. 12–13), the elementary form of power in this layer is 'corrective influence,' which '. . . operates through the use of resources that can serve as punitive and remunerative sanctions that are able to work directly on the interests of subalterns in power relations.' Corrective influence has two sub-types: force (inflicting pain or death, denying food, destroying property by the use of physical strength, weapons, prisons, etc.), and manipulation (for example, advertising, propaganda, price adjustment by the use of money, credit, access to employment). It is through these sub-types of corrective influence that '. . . subalterns can be caused to act or to be prevented from acting by direct restraint or by influence over the conditions under which they make their calculations' (Scott, 2001, p. 13). Power in this layer is effective and economical (Clegg, 1989, p. 18). The result is a space of action between the agents involved, which is characterized by superordination and subordination.

The second layer, power as a relational phenomenon, refers to the fact that (social) power is always exercised in interactions by actors, with relative capabilities, within the context of unintended or even unknown chains of interdependencies. Within these interactions 'persuasive influence' (Scott, 2001, pp. 13–15) is the main elementary form of power. Persuasive influence operates through the offering and acceptance of reasons for acting, and depends mostly on arguments, appeals and beliefs. Where persuasion works through cognitive symbols (ideas and meanings that allow people to define situations), it takes the form of signification. Where it works through value commitments to particular ideas, conditions or competencies, it takes the form of legitimation and/or trust.

Power as a relational phenomenon accentuates relations of autonomy and dependence between agents within an institutional context. Not only the elementary forms of power are then relevant, but also dimensions of intentionality (consciousness), resistance (counteraction), and anticipated reactions. In other words, power relations are more fully developed and (social) power is considered as a catalyser for 'social integration.' Therefore, this layer of power can be studied on the level of networks, configurations or (inter-)organizational practices. The power is 'dispositional' (Clegg, 1989, pp. 2, 83–85) or 'intransitive' (Goehler, 2000, pp. 48–50) in the way that it '. . . constitutes a community as an effective body in the form of a common space of action which is symbolically present.' Agents then act on the basis of shared value conceptions and principles of order, shared symbols, and a common sense of identity in the same common space. Shared cognitive meanings and shared value commitments are based on the kind of (morally and emotionally) appropriate reasons for action that can be offered to others and be regarded as plausible by them (Scott, 2001, p. 13). Bourdieu (1986) has called these resources 'cultural' and 'symbolic capital.' In other words, power in this layer is considered as the collective property of whole configurations of cooperating actors, of the fields of social relations within which particular actors are located (Scott, 2001, p. 9). From a normative perspective, this layer of power represents the human condition as formulated by Hannah Arendt (1958, p. 200): 'Power is what keeps the public realm, the potential space of appearance between acting and speaking men, in existence.' That is why Arendt considered discursively formed power in a positive way; it is a form of 'collective empowerment.' It is worthwhile to note that Foucault tends to stress the negative face of discursively formed power, because as the result of socialization and seduction it is a source of social control and 'discipline' (Scott, 2001, p. 11). This approach introduces a different perspective on power as the 'collective property of whole configurations of cooperating actors' (see above). Disciplinary power is dispersed through all the groups, organizations, and agencies of a society. 'Societies tend to be highly fragmented, forming dispersed 'archipelagos' of localized discursive communities, each of which is the basis of its own specialized forms of power' (Scott, 2001, p.12).

The third layer, power as a structural/systemic phenomenon, refers to the structured asymmetries of resources as a result of specific structures of signification, legitimation, and domination in a certain context and period. In other words, power as a basis for 'systems integration' (Coleman, 1977). This allows us to incorporate also the problem of inclusion and exclusion into the analysis of the balance of power. Power is shaped by, and, in turn, shapes the socially structured and culturally patterned behaviour of groups and the practices of organizations.

To a certain extent a clash of vocabularies seems to exist around the concept of power (Haugaard, 1997, p. 21). It should be emphasized, however, that the layers in the concept of power are distinguished for analytical reasons only. A conception of power is needed that allows us to acknowledge both the influence of actors on the development of policies in networks, configurations and so on, as well as the impact of the structural context in which the actors operate. In this part of the book none of the authors reduces power intentionally and explicitly to either axis of the debate over agency and structure. Instead, it seems that the authors implicitly have followed Clegg in his statement (1989, p. 20) that '. . . power is best approached through a view of more or less complex organized agents engaged in more or less complex organized games.' That brings us to the concept of 'power dynamics.'

Power dynamics can be considered in two ways (see also Figure 1 from the general introduction). First, it reflects the power relations caused by the ever-changing forms of social capital and political participation. Second, power dynamics is the process that produces and reproduces (the circuits of) power as a capacity, power as a relational phenomenon, and power as a structural / systemic phenomenon.

By studying the dynamics of power in rural affairs, it can be expected that some key questions from the *Budapest Declaration* (2002) can be answered, at least provisionally:

- Do the power relations change, actually and/or differently, under conditions of new social capital and political participation in rural areas in Europe? If continuity more than change in power relations reflects the actual situation, why then do the dominated tend to comply instead of to revolt? In other words, how can we explain the reproduction of processes of social and systems integration in the rural areas?

- Who controls and who benefits from the perceived change in power relations in rural development? Is the control in the hands of a (new) local elite? Do local and regional organizations in the rural areas have enough trust in the regime of a deliberate democracy and its procedures of interactive policy-making?
- How do local agents and local networks manage the challenges produced by structural powers at the national, regional (EU) and global (WTO) level?
- Do dynamics of power imply productive rural innovation, particularly in the field of multi-actor and multi-level governance?

Overview of Part I

Several chapters of this book apply the three-layered concept of power to explain the success or failure of specific new policy arrangements and political institutions concerning rural development in general and rural innovation in particular.

From the concept of 'family strategy' it can be explained why farm families are disposed or not disposed to accept the rules of (new) policy-games and (new) policies (i.e., dispositional power). Mauléon argues that the Basque public institutions should be aware of the diversity in the behaviour of family farmers facing the agrarian policies they draw up, and they should take into consideration the domestic group rather than the farm as a unit of analysis. The 'family strategy' concept seems to include also the cumulation of power resources a specific family can mobilize to express its interests in a certain process of policy-making and policy-implementation (episodical power). Finally, the concept of 'family strategy' not only concerns the power resources owned by a specific family, but also the 'embeddedness' of the family – in other words its capacity to operate effectively in its contexts (family related to structural power).

Laschewski and Siebert focus on 'social capital formation in rural East Germany' and emphasize many qualities of post-paternalism. Local structures are rather hierarchical and a few actors monopolize linkages to the rural development system. In fact, these actors, who predominantly were members of the old hierarchy of the large agricultural firms, behave as a local elite. The dynamics of power seem to be based on the ownership of local resources as well as on the centrality in the political-economic arena of some single actors, who can link the local community to external support systems.

According to Hyyrylainen, in the case of Finnish rural development, social capital is a vital power resource in matters of local development and political participation at the local and regional level of governance. The new local partnerships are based on practical commitments to the vitality of the rural areas in which all participants are ready to invest in accordance with their capacities (power resources). They seem to be a counteraction (in the form of pressure as well as protest) to the process of sectoralization and specialization of societies, which has gone too far in relation to the amount of people available in some areas.

This puts another issue back on the agenda. As long as local actors are perceived as 'incapable' and rural areas as 'underdeveloped' they become subjects for political intervention (see also Bock, 2002 for the Netherlands). In Hungary for example, the evolution of expert networks of rural developers is closely connected to an image that constructs the 'rural' as a 'miserable countryside.' It is true that the abolition of the monolithic party state has been followed by a decentralization of power. For example, in Hungary the number of administrative staff in local government offices doubled in the period between 1970 and 2000. In this context a new administrative and economic (local) elite has emerged, because the administration and the practices of rural policies has to be organized according to European standards (Kovach, 2002, pp. 110–113).

At a higher level of abstraction, shifts in power/knowledge relations as expressed by shifts in issues, ideologies, rituals, rules etc., have been discovered through the use of discourse analysis as a research method. The contributions by Lysgård (Norway) and Goverde (EU–Dutch policy relationships) in particular demonstrate that the 'archipelago' of discursive communities is characterized by different discourses concerning rural development as well as by rural policies at different administrative levels of governance. These contributions seem to suggest that discourses and discursive formations function as sources of power, which have a wider scope than the consciousness and the intentions of individual people. The discovery and the narrative of a discourse help to explain why a certain perception of a problem (for example the gendered nature of rural development) becomes dominant and authoritative, while other points of view receive weak support. Because the discursive deliberations give meaning to our world (social environment), it can be explained why a 'discourse,' in fact why 'the existence of multiple interpretative horizons' (Haugaard, 1997, p. 183), has 'constitutive power.' In this perspective all types of policy categories are not only public instruments for realizing public goals, but also signs and symbols that give meaning to, for example, 'rurality' in specific physical areas.

References

Arendt, H. (1958), *The Human Condition*. Chicago and London, The University of Chicago Press.

Bock, B. (2002), *Tegelijkertijd en tussendoor. Gender, plattelandsontwikkeling en interactief beleid*. Wageningen, Wageningen University (PhD thesis).

Bourdieu, P. (1986), The forms of capital. In: John G. Richardson (ed), *Handbook of Theory and Research for the Sociology of Education*. New York: Greenwood Press, pp. 241–58.

Clegg, S.R. (1989), *Frameworks of Power*. London, Sage Publications.

Coleman, J. (1977), Notes on the study of power. In: R.J. Liebert and A.W. Imerskein (eds), *Power, Paradigms and Community Research*. London, Sage Publications.

Faulks, K. (1999), *Political Sociology. A Critical Introduction*. Edinburgh, Edinburgh University Press.

Giddens, A. (1982), Action, structure, power. In: A. Giddens (ed.), *Profiles and Critiques in Social Theory*. London, Macmillan.

Goehler, G. (2000), Constitution and use of power. In: Goverde, H. et al., *Power in Contemporary Politics*. London, Sage, pp .41–58.

Goverde, H. and J. van Tatenhove (2000), Power and policy networks. In: Goverde, Henri, et al., *Power in Contemporary Politics*. London, Sage, pp.96–111.

Goverde, H. et al. (2000), *Power in Contemporary Politics. Theories, Practices, Globalizations*. London, Sage.

Haugaard, M. (1997), *The Constitution of Power. A Theoretical Analysis of Power, Knowledge and Structure*. Manchester, Manchester University Press.

Held, D. (1995), *Democracy and the Global Order*. Cambridge, Polity, Blackwell.

Kovach, I. (2002), Leadership, local power and rural restructuring in Hungary. In: Halfacree, K., I. Kovach and R. Woodward (eds), *Leadership and Local Power in Contemporary Rural Europe*. Aldershot, Ashgate, pp. 91–122.

Little, J. (2002), *Gender and Rural Geography: Identity, Sexuality and Power in the Countryside*. Englewood Cliffs, New Jersey, Prentice Hall.

Lovendus, J. and S. Stephenson (1999), *Women in Decision Making*. Report on existing research in the European Union. Luxembourg, EC, DG V, Office for Official Publications of the European Communities.

Scott, J. (2001), *Power*. Cambridge, Polity Press, Blackwell.

Shortall, S. (2002), Gendered argricultural and rural restructuring: a case study of Nothern Ireland. *Sociologia Ruralis* 42 (2), pp. 160–176.

Chapter 3

Social Capital Formation in Rural East Germany

Lutz Laschewski and Rosemarie Siebert

In an earlier publication (Siebert and Laschewski, 2001), we addressed the missing linkage of newly established state institutions to the rural society in East Germany as a fundamental problem for rural development. In that article, we focused on the analysis of institutional change. What has been missing, still, is an attempt to develop a general understanding of the nature of rural society in East Germany more than ten years after the unification. In this paper, we will draw on the idea of paternalism, which we outlined previously (Laschewski and Siebert, 2001), to characterize the rural society. Here, going beyond this earlier writing, we want to relate this analysis more closely to current discourses in rural studies. However, when we, as we do, argue that social capital formation is central for rural development in East Germany, we also have to ask how it can be done and which role the state can play. For the case of East Germany, it can be said that the state is currently rather a part of the problem than the solution.

Halfacree, Kovach and Woodward (2002) suggest that the role of local elites and local power networks is crucial understanding the practical outcomes of rural development policies. In the same publication the authors argue that a research methodology, such as the actor-network theory, seems to be an appropriate approach to capture the addressed issues. One of the most prominent concepts, which has been linked to networks and networks theory in general and which has also become a political catchword (not only) in the rural development debate, is social capital. By very broad definition social capital can, first, be understood as the capability of social groups to act collectively. Second, social capital is also described as an individual resource resulting from somebody's social position within a particular field of action.

Network theory plays an important role for both definitions of social capital. It also relates social capital to the concept of power. First, social capital as an individual source of power derives from one's social position within in a network. Centrality of a social position is among the most common measures to describe positional power. However, the most central position within a network may not always be the most powerful position (Jansen, 1999, pp. 121 ff.). A second, and for our analysis even more important, result of network analysis is that social capital derives also from those positions that link different networks. Such weak (social) ties (Granovetter, 1973), which bridge 'structural holes' (Burt, 1992) constitute social capital too.

From a network theory perspective, individual social capital and collective social capital are not necessarily exclusive. The first argument to prove that is that the network position can also be described for a group of actors of even networks themselves. Second, some forms of individual social capital have clearly been identified to strengthen collective social capital too. For instance, the structural autonomy of actors, which derives from the ability to bridge structural holes at so-called 'cutpoints' (Burt, 1992), is an individual form of social capital. As social capital theorists argue, the density of social linkages between networks is a form of collective social capital which is based on such actors and brings about trust and norms of reciprocity on which successful co-operation is based (Putnam, 1993, 173). In this sense collective social capital goes along with structural autonomy of many people.

Many of these issues addressed by social capital theorists have been put forward by rural sociologist under the label of 'endogenous' development (e.g., van der Ploeg and Long, 1994; van der Ploeg and van Dijk, 1995), and in German speaking countries as 'autonomous' development (Pongratz and Kreil, 1991). The main idea of this debate was to draw on internal community resources for rural development rather than to rely on external support. This general idea has been subject to further elaboration in line with Woolcock's (1998) argument that social capital is not only the ability to draw on intra-community resources, but also related to the existence of linkages to extra-community networks. A similar idea was brought forward by Lowe et al. (1995). They say that it might be the case that peripheral regions may not be able to generate development from within. Therefore, they plea to go beyond endogenous growth models and 'recognize and indeed celebrate interconnections between areas and between networks' (Lowe et al., 1995, p. 104). They stress the issues of participation and power within the development process. Empirical findings appear to confirm this view (Terluin, 2001).

The crucial aspect of endogenous growth is that it draws on social capital as the ability to mobilize the potential of an existing 'active society' that will generate entrepreneurs, and self-help institutions from within. In the recent debate, more and more scientists have stressed the diversity of local potential. This leads us to two core issues: First, the question is how can social capital be built? Second, what role does the (local) state have to play in this process?

Putnam (1993) described social capital formation as a centuries lasting process. In this sense, social capital is a cultural tradition and has to be treated as an exogenous variable (Paldam and Svensen, 2000, p. 347). It may be argued that the low rate of social capital formation is due to little rational consideration of its development effects. Currently, it appears to be an open question in how far social capital formation can be influenced at all. Often it comes about as a positive externality of activities aiming at another purpose. An optimistic view is to invest into institution building strategies and attempts to encourage co-operative movements within development policies. However, experiences indicate the huge difficulties and the slow progress made (ibid.).

One basic problem is seen in the role of external and, in particular, state agencies to enforce participation. Many scientists argue that there is little or no opportunity for the state to contribute to social capital formation. Indeed, the basic idea of social capital refers to voluntary sector activities, and self-enforcement rather than third party enforcement. Attempts by third parties – as public authorities – to enforce social capital may thus be counterproductive (Paldam and Svendsen, 2000, p. 366). Warner (1999) questions this view. Her argument is that it depends on the local context, i.e. the local power structures, which effects will result of policy influences. Referring to Woolcooks' (1998) relationship between linkages, she argues that the structure of autonomy and linkages determines how communities respond to external influences. It is, in particular, in communities characterized by hierarchical social capital and weak or patronage government systems where outside support is required. In this context, devolution may be counterproductive. The idea of hierarchical social capital itself draws on the concept of paternalism (Schuman and Anderson, 1999). Paternalism describes a particular social setting that is rather common in rural places. We think that its basic characteristics are important for the understanding of rural East Germany (and probably other Central and Eastern European countries too). Therefore, we outline the basic characteristics of paternalistic local structures, and illustrate some experiences with its transformation under conditions of rapid economic change.

Paternalism, its Nature and Transformation

The concept of paternalism has been used to describe a traditional form of authority (In the following, we refer, in particular, to Ackers and Black (1991) and Newby (1978). Basic elements are the differentiation between classes, concentration of power and identification of the subordinate class with the members of the dominant class. The particular context of paternalistic systems is the industrial society starting in the late 19th century. The basic characteristic of a local paternalistic system is the existence of large firm in a relatively isolated geographical context. Due to its domination of the local labour market, the firm is also dominating all other aspects of community life. However, the second feature of paternalism is an ethic of social responsibility among the dominating class. Due to this moral orientation the elite tends to provide and support a wide range of social services and facilities, such as housing, health care and schooling.

Over time more and more aspects of local life are structured around the dominant firm, either symbolically, e.g., by naming public places after the firm or their owners, or physically, for instance by developing housing areas close to the plant. The close interrelation of the firm with the local community leads to the situation that the history of many families is closely connected with the history of the firm. The recruitment of new staff is, therefore, typically oriented towards such family networks, and the identification of the employees with the firm is usually very high. Typically, we find a clear gender division of labour and a strong male domination within such a system. Private and workplace relationships overlap due to family employment, the firm's influence on community life and the existence of occupational housing.

Generally, paternalism is seen as being connected with family ownership and the idea of the owner-managed firm. However, under western style capitalism many family firms have been transformed into companies with a diversified owner structure. Furthermore, there evolved quasi-paternalistic systems in nationalized industries (e.g. coal mining in Britain). Likewise, such systems developed under socialist conditions in Central and Eastern Europe. Here, the state as a body and his representatives in persona are in the position of the dominating class of the owners.

There are different opinions on the degree of unionization in paternalistic systems within Western economies. While, for instance in Britain industry's most well known paternalistic employers have long encouraged and well-organised trade unions, in the US, paternalism is widely described as a structure that fought against unionism. However, paternalism produces paternalistic trade unions. Those tend to be small and parochial locally based organizations.

In many cases, paternalism is understood as reminiscence of early industrial times. It is argued that industrial paternalism has vanished because of bureaucratization and the withdrawal of family owners from management. However, as argued above, forms of paternalism can also be identified without the existence of the family owner domination. Paternalism has also been identified in non-industrial sectors, such as agriculture. Hence, what appears to be more significant are social and economic changes, which undermine paternalism in western societies. On the one hand, the economic base of paternalism, which is usually connected with the primary sector or traditional industries, has come under immense pressure because of forced global competition. On the other hand, the geographical isolation has decreased in line with increasing mobility, and as a consequence geographically expanding labour markets.

In an ideal type situation, we would expect that the transformation of paternalistic local structures would allow for more independent local policies, a smaller influence of the dominating firm on local social life, and scope for new entrepreneurial activities. In a short, the ideal development is characterized by a stronger differentiation of private, public and voluntary sector institutions and the involvement of a wider range of local actors.

However, in many cases the experiences are rather different. First, we find that the development of a small business sector, which is able to compensate for job losses and constitutes the base for an active society, is rather the exception than the rule. The arguments for that are manifold (e.g., Rees and Thomas, 1991; Laschewski, 2000). Some stress explicitly the subjective barriers of workers to become entrepreneurs. It is no doubt that it means a huge psychological effort to overcome the habitus of a 'deferential worker' (Newby, 1978), and develop entrepreneurial spirit. However, the structural barriers are also huge. Many workers with similar knowledge and experiences start to seek for opportunities in an environment that does not offer many. Under such conditions, long-distance commuting or migration often offers a more realistic chance for many, while the new entrepreneurs have often been the middle managers or specialists of the former dominant firm before. As a social consequence, we observe the evolution of new middle classes and also of a large social group of losers (see, e.g., Schuman and Anderson, 1999).

Due to the weak development of the business sector the old paternalistic firms' (or its successors') position remains, despite economic decline, rather dominant in the local labour market. Moreover, they have still control of important local resources, such as land, local housing estates etc. Nevertheless, those firms tend to withdraw from social activities. Therefore, voluntary activities face severe constraints due to lack of support.

In the following, it is argued that local rural structures in the former GDR can be understood as quasi-paternalistic structures. Therefore, the experiences made in East Germany are to a large extent comparable to the development in other rural areas where paternalistic local structures came under economic pressure.

Rural Structures in the GDR

Under socialism, a system of large-firm paternalism evolved, which was specific for rural areas in the GDR since the 1970s. This system was based on, first, the concentration and collectivization of small businesses in large companies organized either as co-operatives or state firms (for these processes in agriculture see Laschewski, 1998, for industry and service see Albach and Witt, 1993). Secondly, it was based on the regulation of the labour market and also the housing market within planned economy. Due to the artificial isolation based on regulation, the type of rural paternalism, which will be described below, also existed in accessible locations. Finally, it was based on monopolistic domination of the socialist unity party, together with a caring ethic within the socialistic ideology.

Rural employment was largely based on agriculture and industries. The latter were located in larger rural towns. Locally there may have been other large employers, such as the army or tourist resorts. However, for most villagers there were two main employment options: either being employed by the local agricultural (co-operative or state-owned) firm or to commute to the next town. In particular in the remoter rural areas, agricultural firms employed up to a third of the labour (Rodewald and Siebert, 1995). However, only about 60 percent of the jobs in such firms were

agricultural jobs. Beyond those, there was a wide range of activities such as administrative and social services, building and construction, food processing, technical services and transportation (BMELF, 1991; Großkopf and Kappelmann, 1994).

Agricultural firms provided a wide range of services for the community (Parade,
1998). Within the rural communities a transfer of functions from the municipality to
the agricultural firm took place, which in many cases was accompanied by a power
transfer from communal bodies to the firm. Frequently, chairmen of agricultural
firms were more influential than local mayors were (Zierold, 1997). Owing to financial restrictions of the municipality and the general supply shortages under socialism,
agricultural firms were the central investors in local infrastructure (e.g., local roads,
village halls) and suppliers of a variety of local services. As villages were greatly dependent on agricultural firms, the infrastructure was mainly in the interest of the
firm. Agriculture firms were the main user of roads, they needed halls for their own
assemblies and built houses to attract workers.

For social and cultural activities too, use had to be made of facilities provided
by agricultural firms. Social and cultural funds were sponsored by firms, which also
provided transportation or other services. In some cases, clubs (e.g. horse riding)
were founded and supported by the firms. Often, agricultural firms initiated cultural
events for the community (Zierold, 1997), and social events for pensioners. Further
social services that were usually provided by agricultural firms, were child nurseries,
canteens, and holiday homes and camps. Since in almost every village family somebody was employed in an agricultural firm, a firm event was almost always a village
event too.

Unfortunately, there has been little research on personal relationships among villagers and agricultural firms. The position of the chairman of the firm had, as already
mentioned, been fairly strong. However, he or she was usually a party member, and
therefore bound to party decisions. On the other hand, chairmen were also mostly
integrated into dense family networks within the community. Studies made after 1990
indicated that the 'old peasantry' continued to play a distinctive role within many
communities (Brauer et al., 1996; Laschewski, 1998). An important part of the population came as refugee from the former Eastern parts of Germany, and has been settled
in rural areas after the Second World War, and mostly started with huge difficulties as
'new peasants' after the land reform of 1945.

There has also been little research on the participatory practice within firms and
villages. Formally, democracy, particularly within agricultural co-operatives, was very
strong. However, at the firm level, there was no realistic chance to influence economic
plans made by state authorities. Representation through unions was also very weak
and the labour regime comparatively rigid. On the other hand, the board consisted
of about fifteen to twenty people out of the staff, some of them nominated as women
or youth representatives. Therefore, it is quite likely that social issues played an important role. There was also a wide range of committees, which were at least formally
open to everybody and not limited to the employees of the co-operatives. Although

the power of committees might have been rather confined, it guaranteed the involvement of many. The integrative role of agricultural firms and their undeniable social contributions may explain the high degree of identification by employees, which was expressed in 1990 (Hubatsch et al., 1991).

Transformation Problems in Rural East Germany After the Unification

After the unification the rural regime in East Germany came under pressure because of a dramatic agricultural and industrial decline. Increasing mobility due to the liberalization of labour and housing markets and the increasing availability of cars. With the integration of East Germany into the common agricultural market the highly subsidized East German agriculture had to increase productivity. At the same time institutional reform was imposed on agricultural co-operatives while state farms were abolished (Wiegand, 1994; Laschewski, 1998). Within in two years about 80 percent of the labour force was made redundant. After ten years multiple forms of agricultural firms have emerged from producer co-operatives, stakeholder societies, and limited companies to comparatively large family farms and any type of small-holdings. Successors in a variety of legal forms dominate in many regions. However, those agricultural firms, which were at the core of village life, now can hardly offer sufficient employment opportunities let alone that they can provide support for the local community beyond their economic activities.

One part of the legacy of paternalism in East Germany is the lack of encouragement of entrepreneurial initiative in other than the agricultural sector. People, in particular in northern Germany, respond to these problems with emigration rather than initiative. There is a lack of innovation and actors that take risk. A central problem is youth emigration (BBR, 2000). Few job offers and insufficient training opportunities are core problems that cause significant emigration among young people between 18 to 25. There is a danger of developing a 'culture of emigration,' which affects the development potential of rural areas (Siebert, 1999). The remaining, partly less mobile population, cannot make up for these emigration losses. It is rather likely that providing services for elderly people may become a problem in the near future.

Agricultural decline has also affected the provision of services in rural villages. This led to a reduction in many services. In a community study, Rodewald (1994) described the situation as follows: 'All social and cultural services (health station, pub, local shop) have meanwhile been closed. Institutions such as child nursery, day home for schoolchildren and classrooms are abandoned.' Communities have to find their own, self-reliant status as a functioning unit independently from agriculture (Herrenknecht,1995). Most communities are financially dependent on external funds. The room for action of local policy makers is so narrow that the potentially existing rights for self-administration cannot be successfully used (Bußmann, 1998). However, there have been significant improvements of the physical infrastructure.

Finally, the voluntary sector has also been significantly transformed. Before unification local clubs were in many cases voluntary organizations, but local groups of mass organizations were either affiliated to the socialist party or of regional associations under state control. Because of that they received comparatively stable financial and personal support by the state, but also from local agricultural firms (Berking, 1995; Hainz, 1998). With unification, the system of state- and firm-sponsored culture and sport suddenly vanished. Clubs, leisure and interest groups, which had existed for decades, closed in numbers.

Generally, there is a decreasing interest in voluntary activities in rural communities. On the one hand, local clubs that have survived, often suffer from a lack of interest. A typical example is the voluntary fire brigade (Rodewald, 1994). On the other hand, only few clubs are founded. A further indicator of weak civil engagement is the poor uptake of AGENDA 21 initiatives in East Germany (Witzschel, 1999; Kolloge, 2000).

In a way it is surprising that neither existing self-help networks nor the democracy movement have built the base for bottom-up initiatives. According to Brauer (2001, p. 62), it is not a general lack of local initiative, but the way how local actors are treated as subjects rather than actors in the planning and development system and live by chance in the planning area. Under such circumstances, approaches of 'civic participation' are to inform and teach local people rather than encourage bottom-up initiatives. Parallels to political agitation and attempts to mobilize local actors for public activities of socialist times are obvious (ibid., Laschewski and Siebert, 2001, p. 40). Bruckmeier (2000) argues that LEADER II in East Germany has served as a conventional rural development tool rather than a measure to integrate independent projects. The village renewal programme has been quite successful as a measure to improve the local infrastructure, but it has not been a successful tool in strengthening local participation. It appears that the sheer amount of development programmes (Brauer, 2001) and the tendency to favour large-scale projects (Beetz, 2001, p. 83) are core problems which seem to stabilize the involvement of the few rather than the participation of many. Furthermore, it is an irony that the agricultural policy designed to support family farms stabilized the agricultural sector, dominated by large farms that are rather disconnected from the rest of the rural economy (Laschewski and Siebert, 2001). In all cases, exclusive elements are the bureaucratization and professionalization of planning procedures that, on the one hand, form a barrier for 'real' participation. They encourage, on the other hand, the development and maintenance of networks of professionals and development experts, which are only loosely coupled to the majority of local actors.

However, further arguments for these trends derive from existential problems of the local population, the psychologically negative impact of unemployment, time constraints because of the increased mobility (long-distance commuting), and from new leisure-time opportunities. There is also a priority of improving the individual materialistic conditions (e.g., building or refurbishing a house) among those, who are potential core actors (Rodewald, 1994). Finally, there is a perception that social engagement is not publicly accepted and honoured, but is in many cases even seen as attempts to 'search for individual profits' or as 'support and stabilization of the socialist system' (Rodewald, 1994; Hainz, 1998). Brauer (2001, p. 63) argues that the call for self-reliance and egoism immediately after unification has also contributed to the fact that collective action has got a negative notion.

Social Capital Formation and Rural Development Policies – Some Conclusions

Rural East Germany shows many characteristics of post-paternalism. From a network perspective, local structures are rather hierarchical and linkages to the rural development system are monopolized by few actors, who predominantly were members of the old hierarchy of large agricultural firms. Power, therefore, is based not only the concentrated control over (which is not necessarily identical to ownership) of local resources, but even more on the centrality of single actors, which link the local community to external support systems. The majority of local social groups is not directly linked to the rural development system.

According to Warner (1999) such a situation calls for state intervention. However, there is not at all a lack of state intervention and development programmes in East Germany. Therefore, the rural development problem is not only the result of hierarchical local structures and a lack of horizontal forms of social capital, but it is also re-enforced by the existing planning and development system. As already mentioned, both planning procedures and the project scope appear to be major barriers for the integration of the marginalized majority into the decision-making process. However, as Brauer (2001) illustrates, it is also the fact that local actors are only entitled to choose between given programmes rather than to develop their own projects, which thus frustrates civic engagement. Brauer (2001, p. 64) seeks the reasons for this deficit in the planning system and the appearance of external (West German) planners in the countryside. Instead, local policy researchers indicate that technocratic-paternalistic thinking among the local administration appears to be in line with the socialist past (Thumfahrt, 2002, p. 657). However, it is obvious that local actors would also bring in a different understanding what rural development should be about. Local actors often address community rather than economic development issues (Brauer, 2001, p. 60). Social capital, therefore, encompasses a quite paradoxical element of the 'usefulness of the useless' (Laschewski and Siebert, 2001, p. 40).

This puts another issue back to the agenda, which also shows the limitations of network theory: the problem of social meaning and culture (see Emirbayer and Goodwin, 1994). As long as local actors are perceived as 'incapable' and rural areas as 'undeveloped' they become subjects for political intervention. As Csite (1998, 2001) illustrated for Hungary the evolution of expert networks of rural developers is closely connected to a view, which constructs the 'rural' as a 'miserable countryside.' But the causal relations between network evolution and social ideas are not always clear. However, this argument shows that the development of horizontal social capital will only be successful in connection with a different understanding of rural development, which recognizes the dimension of democratization, civic engagement and local participation as development objectives and not only as another additional development factor to increase economic growth.

Such a perspective would also allow addressing the problem of social exclusion. Rural research mostly focuses on the economic development and rarely on social integration. As some network theories argue, it depends on the nature of the social problem which form of social capital is more important (Jansen, 1999, p. 256). While social capital based on strong ties is more important for social integration, weak ties are seen as more related to performance. However, only little is known about family support networks and the role of strong ties in rural East Germany, in general.

References

Ackers, P. and J. Black (1991), Paternalist capitalism: an organization culture in transition. In: Cross, M. and G. Payne (eds), *Work and the Enterprise Culture*. London, Falmer, pp. 30–56.

Albach, H. and P. Witt (1993), *Transformationsprozesse in ehemals Volkseigenen Betrieben*. Stuttgart.

Bauerkämper, A. (1994), Von der Bodenreform zur Kollektivierung – Zum Wandel der ländlichen Gesellschaft in der Sowjetischen Besatzungszone Deutschlands und der DDR 1945–1952. In: Kaelbe, H., J. Kocka und H. Zwahr, *Sozialgeschichte der DDR*. Stuttgart, Klett Cotta.

BBR (2000), *Raumordnungbericht 2000 (Entwurf)*. Bonn.

Beetz, S. (2001), Woher die Menschen und wohin mit dem Land? *Berliner Debatte INITIAL* 12 (6), pp. 77–86.

Berking, H. (1995), Das Leben geht weiter. Politik und Alltag in einem deutschen Dorf. *Soziale Welt* 46, pp. 342–353.

BMELF (ed.) (1991, 1994, 1998), *Agrarbericht der Bundesregierung*. Bonn, BMELF.

Brauer, K., A. Willisch and F. Ernst (1996), Intergenerationelle Beziehungen, Lebenslaufperspek-tiven und Familie im Spannungsfeld von Kollektivierung und Transformation. In: Clausen, Lars (ed.) *Gesellschaften im Umbruch – Verhandlungen des 27. Kongresses der Deutschen Gesellschaft für Soziologie in Halle an der Saale 1995*. Frankfurt a.M./New York, Campus.

Brauer, K. (2001), Unsere Lösung – Ihr Problem. *Berliner Debatte INITIAL* 12 (6), pp. 52–64.

Burt, R.S. (1992), *Structural Holes. The Social Structure of Competition*. Cambridge/Mass.:Harvard University Press.

Bußmann, E. (1998), Dorfbewohner und Kommunalpolitik. Eine Vergleichende Untersuchung in 14 Dörfern der Bundesrepublik Deutschland unter besonderer Berücksichtigung der länderspezifischen Gemeindeordnungen und der Verwaltungsstrukturen. In *FAA*, Bd. 309, Bonn.

Csite, A. (1998), Constructing the miserable countryside in Hungary in the 1990s. In Granberg, L. and I. Kovách (eds), *Actors of the Changing European Countryside*. Budapest: Institute for Political Science.

Csite, A. (2001), Europeanising rural Hungary – rural policy networks and policy representations of the countryside in Hungary in the 1990s. In Torvey, Hilary and Michael Blanc (eds), *Food, Nature, and Society – Rural Life in Late Modernity.* Aldershot: Ashgate, pp. 253–273.

Emirbayer, M. and J. Goodwin (1994), Network analysis, culture, and the problem of agency. *American Journal of Sociology* 99, pp. 1411 –1454.

Granovetter, M. (1985), Economic action and social structure. *American Journal of Sociology,* pp. 481–510.

Großkopf, W. and K.-H. Kappelmann (1992), Bedeutung der Nebenbetriebe der LPGen für die Entwicklungschancen im ländlichen Raum. In: Schmitt, G. and S. Tangermann (eds) *Internationale Agrarpolitik und Entwicklung der Weltagrarwirtschaft.* Schriften der Gesellschaft für Wirtschafts- und Sozialwissenschaften des Landbaues e.V., Band 28, Münster, Landwirtschaftsverlag.

Halfacree K., I. Kovach and R. Woodward (2002), Conclusions. In Halfacree K., I. Kovach and R. Woodward (eds), *Leadership and Local Power in European Rural Development.* Aldershot, Ashgate.

Hainz, M. (1998), Dörfliches Sozialleben im Spannungsfeld der Individualisierung. FAA, Bd. 311, Bonn.

Herrenknecht, A. (1995), Der Riß durch die Dörfer – Innere Umbrüche in den Dörfern der neuen Bundesländer. In Agrarsoziale Gesellschaft, *Dorf- und Regionalentwicklung in den neuen Bundesländern – Beiträge aus der Praxis,* Kleine Reihe Nr. 54. Göttingen, pp. 50–64.

Hubatsch, K., K. Krambach, J. Müller, R. Siebert and O. Vogel (1991), *Genossenschaftsbauern – Existenzformen Lebensweise im Umbruch.* Berlin, Forschungsbericht.

Jansen, D. (1999), *Einführung in die Netzwerkanalyse: Grundlagen, Methoden, Anwendungen.* Opladen, Leske + Buderich.

Kolloge, S. (2000), *Die Wirkung der Agenda 21, Eine institutionetheoretische Wirkungsanalyse am Beispiel von sechs Kommunen im ländlichen Raum Großbritanniens und Deutschlands.* Berlin, unveröffentlichte PhD Dissertation.

Laschewski, L. (1998), *Von der LPG zur Agrargenossenschaft. Untersuchungen zur Transformation genossenschaftlich organisierter Agrarunternehmen in Ostdeutschland.* Berlin.

Laschewski, L. (2000), *The Formation and Transformation of Business Co-operation – Some Preliminary Observations from the North East of England.* Centre for Rural Economy Working Paper Series 46, Newcastle upon Tyne: Centre for Rural Economy.

Laschewski, L. and R. Siebert (2001), Effiziente Agrarwirtschaft und arme ländliche Ökonomie? *Berliner Debatte* INITIAL 12 (6), pp 31–42.

Newby, H. (1977), *The Deferential Worker: A Study of Farm Workers in East Anglia.* London, Allen Lane Penguin Books.

Paldam, M. and T. Svendsen (2000), An essay on social capital: looking for the fire behind the smoke. *European Journal of Political Economy* 16, pp. 339–366.

Ploeg, J.D. van der and A. Long (eds) (1994), *Born from Within: Practice and Perspectives of Endogenous Rural Development.* Assen, Van Gorcum.

Ploeg, J.D. van der and G. van Dijk (eds) (1995), *Beyond Modernization – The Impact of Endogenous Rural Development.* Assen, Van Gorcum.

Pongratz, H. and M. Kreil (1991), Möglichkeiten einer eigenständigen Regionalentwicklung. *Zeitschrift für Agrargeschichte und Agrarsoziologie* 39, Heft 91, pp. 91–111.

Putnam (1993), *Making Democracy Work: Civic Traditions in Modern Italy.* Princeton, Princeton University Press.

Rees, G. and M. Thomas (1991), From coalminers to entrepreneurs? A case study in the sociology of re-industrialization. Cross, M. and G. Payne (eds), *Work and the Enterprise Culture.* London, Falmer, pp. 57–78.

Rodewald, B. (1994), Glasow – ein Dorf im Schatten der Grenze. In FAA, *Ländliche Lebensverhältnisse im Wandel 1952, 1972 and 1992*, Bonn, pp. 443–468.

Rodewald, B. and R. Siebert (1995), *Arbeitsmarkt und Mobilität. Eine empirische Analyse von dimensionen räumlicher Immobilität und Mobilität im Kontext regionaler Arbeitsmarktentwicklungen im ländlichen Raum Ostdeutschlands.* Müncheberg, Forschungsbericht.

Schuman, M.D., and C. Anderson (1999), The dark side of the force: a case study of restructuring and social capital. *Rural Sociology* 64 (3), pp. 351–372.

Siebert, R. (1999), Wandel ländlicher Räume – soziale Rahmenbedingungen für die Landwirtschaft. In *Aktionsbündnis ländlicher Raum: Ländliche Räume, Landschaft und Landwirtschaft 2010*, DLG Arbeitsunterlagen, Frankfurt a. Main, pp. 7–15.

Siebert, R. and L. Laschewski (2001), Becoming a part of the union – Changing rurality in East Germany. In Torvey, Hilary and Michael Blanc (eds), *Food, Nature, and Society – Rural Life in Late Modernity.* Aldershot, Ashgate, pp. 235–252.

Slee, B. (1994), Theoretical aspects of the study of endogenous development. In J.D. van der Ploeg and A. Long (eds), *Born from Within.* Assen, Van Gorcum, pp. 184–194.

Terluin, I.J. (2001), *Rural Regions in the EU – Exploring Differences in Economic Development.* Utrecht/Groningen, Nederlandse Geografische Studies 289, Rijksuniversiteit Groningen.

Thumfart, A. (2002), *Die politische Integration Ostdeutschlands.* Frankfurt a.M., Edition Suhrkamp.

Warner, M. (1999), Social capital construction and the role of the local state. *Rural Sociology* 64 (3), pp. 373–393.

Witzschel, K. (1999), *Lokale Agenda 21.* Diploma thesis, Potsdam.

Woolcook, M. (1998), Social capital and economic development: toward a theoretical synthesis and policy framework. *Theory and Society* 27, pp. 151–208.

Zierold, K (1997), Veränderungen von Lebenslagen in ländlichen Räumen der neuen Bundesländer. In A. Becker (ed.) *Regionale Strukturen im Wandel.* Opladen, Leske & Budrich, pp. 501–567.

Chapter 4

Family Strategies and Farming Changes: The Case of Family Farming in the Basque Country

José Ramón Mauleón

Public institutions establish agrarian and rural policies aimed at improving the economic conditions of farmers. Through these policies farmers may receive funds to modernize their farms, to move from traditional agricultural products to new ones, to start non-agricultural economic activities on the farm, etc. These measures regulate the conditions that must be met by the candidates, the productive changes they must introduce, and the amount granted by the institutions to those eligible. With these measures, the institutions exercise their power influencing the decision making of farmers. Thus, the conditions established for obtaining economic support lead to the introduction of criteria differentiating between farmers, and the definition of the changes to be introduced indicate the model of agriculture sought by the institutions.

Nevertheless, there are some farmers who choose not to become engaged in certain schemes in spite of meeting the conditions required. This fact suggests the existence of a differential farming behaviour with regard to agrarian policies. The aim of this chapter is to try to determine the reasons why not all potential beneficiaries decide to apply for a particular agrarian scheme. In order to achieve this aim, we will need to analyse how the internal dynamic of the family interacts with the external context where it is located. The hypothesis put forward is that the family takes a decision on its future as a family unit considering both internal aspects and external factors, and this decision will condition the productive changes to be introduced on the farm.

In order to test this hypothesis, a group of farmers sharing common characteristics will be analysed. The research will focus on farmers who specialize in the same area of production (dairy products), located in the same context (the Basque Country, Spain), and working on family farms (those which do not employ salaried workers). It is quite a homogenous group since their farming activities are regulated by the same constraints deriving from the Milk Quota programme, and they also share the new economic possibilities offered by the Basque institutional schemes. The research has analysed the productive changes introduced on these farms and the reasons for these changes. We have tried to understand the productive trajectory of the farms within the context of where they are located – what we call the dairy agrifood system – and the decisions taken on family reproduction – what we call *family strategies*.[1]

The information that has been gathered combines quantitative as well as qualitative material. Quantitative data comes from a survey carried out among 308 dairy farmers in 1985. Qualitative information proceeds from eight focus groups that were organized in 1992 involving different types of families. Although the survey data is quite old, it does not invalidate the results, since our objective is not to understand present changes but whether a relationship exists between productive change and the type of family strategy.

Family Farming and Family Strategies

One of the more comprehensive definitions of family farming is that given by Harriet Friedmann (1978, 1980). She defines a farm as a *form of production* constituted by the characteristics of the farm itself and the socio-economic context in which it is located. A family farm may be defined by three characteristics: it only employs family labour, it is located in a capitalist social formation, and it is able to reproduce itself when the final income covers the producers' needs and maintains, or replaces, the means of production.

The definition proposed by Friedmann is of considerable interest because it allows us to take into account two important dimensions of farming: the political and economic context in which farms operate, and their logic of functioning. By allowing room for the *macro context* we are forced to consider the farm as part of the whole society. This permits the inclusion of *external* aspects that are decisive for the farm's performance and progress, such as the model of agricultural development followed by agrarian policies, or relationships between agribusiness and the farmers. Since this definition also refers to the conditions for reproduction, it includes its logic of functioning, that is to say, the internal dynamics of the farm and family. Friedmann's theoretical approach has not gone unnoticed amongst social scientists. There are a number of authors who have made use of it in their research, or who have reformulated it or criticized it (Goodman and Redclift, 1985; Whatmore et al., 1986; Mauleón, 1997).

Friedmann's contribution provides an adequate characterization of family farming, but it does not permit an understanding – it was not her purpose – of the differences in productive changes introduced on the farm. Nevertheless, other authors have proposed a variety of criteria for understanding this diversity. Some of the most popular criteria are: the size of the holding (Nooij and Somers, 1986); the development of the family cycle or the natural history of the family (Galeski, 1977; Nalson, 1968; Chayanov, 1985); the role played by women in family farming (Bouquet, 1982; García Ramón, 1990; Evans and Ilbery, 1996); the level of the farm's commoditization (Ploeg, 1985); the external relations and internal differences (due to gender and age relations) (Whatmore et al., 1987); or the work dedicated to the farm and the ability of farm income to reproduce the farm and the family (Djurfeldt and Waldenstrom, 1996).

The proposal offered in our research differs from the previous criteria and is an attempt to elaborate a typology of family farming from what Cardesín's calls the *family's reproductive project*, that is, whether or not a potential successor decides to continue with the farm as a way of living. This explanation has a background in the agrarian social sciences (Greenwood, 1976; Potter and Lobley, 1992; Marsden, 1989). Nonetheless, while other authors consider that the decision to emigrate or to continue on the farm constitutes a mere *potentiality*, or one of the aspects that can influence the introduction of changes, in our view, it is an event of the first order that will affect farming practices.

Our view raises the need to distinguish between the farm as a production unit, the family as a reproductive unit, and the household as an entity that embraces both (Cardesín, 1992, pp. 47–9). Productive changes are there to serve the reproductive project of the family group. Individual and family group actions consist of adapting their efforts to the opportunities and limitations that arise in ensuring the family's survival during its life cycle. This way of understanding family farming is part of the family strategies approach, a theoretical outline that is gaining importance nowadays.[2] Thus, it is our hypotheses that the family, a sociological variable, emerges as the independent variable capable of explaining innovations on the farm.

We are dealing with a criterion that does not invalidate or exclude others that are mentioned. On the contrary, it brings them together by considering the family as a unit where decisions are taken (not without conflicts), and which embraces members of the family (whether a potential successor decides to continue with the farm), the family as a whole (whether the members of the family – especially the mother – show sufficient support for a common family plan), the farm (productive characteristics of the farm), and external factors (agricultural prices and policies, or possibilities of finding an off-farm job). Previous criteria proposed by other authors concerning the characteristics of the farm or the family (women's role, successor, etc.) emphasize only partial aspects, but fail to offer a more global perspective of the family group. And criteria focused on aspects related to external aspects do not take into account the possibility that family farms may react differently in spite of a common external context.

In order to confirm that innovation on the farm depends on the strategy adopted by the family, a typology of six families was drawn up. Three variables were used to build up that typology: the age and level of commitment to farming of the *central person*, and whether or not there is a successor. By *central person* we refer to the family member most engaged in the farm; in most cases it refers to the father. The presence of a successor is relevant since this shows whether the family has agreed to some extent that an offspring will reproduce a new family on the basis of the farm's income. The next variable, the age of the *central person* on the farm, has been grouped in three intervals: less than 49 years old because it is still too early to know if an offspring will take over the farm; from 50 to 64 years old because a successor may exist; and over 65 years old because after that age there is not a successor (otherwise the *central person* would be the offspring), and because family reproduction depends to some extent on a pension. Finally, the commitment to farming of the *central person* is also important for understanding present family reproduction, since farmers engaged in earning an off-farm income may reproduce the family and even the farm activity through the incomes generated by external employment. By combining these variables, the six different types of families used in this research emerged (see Table 4.1).[3]

Table 4.1 Types of families

Age of central person	Commitment of central person	Successor
1. Aged family		
2. Adult family	Exclusive commitment	With successor
3. Adult family	Exclusive commitment	Without successor
4. Adult family	Partial commitment	Without successor
5. Young family	Exclusive commitment	
6. Young family	Partial commitment	

The Dairy Agrifood System in the Basque Country

The Basque Country is located in the North of Spain. It has a population of around 2,1 million inhabitants concentrated in the capital cities of its three provinces. Sixty-eight percent of the population resides in the metropolitan areas of Bilbao, San Sebastian and Vitoria. The services sector is the most important economic activity. In year 2000, 59.7 percent of the employed population, and 59.8 percent of the Gross Domestic Product (GDP), proceeded from the services sector. The agro-fishery sector, on the contrary, only represented 2.3 percent of employment and 1.8 percent of the GDP (EUSTAT, 2001, pp. 61, 211).

Dairy production is one of the most important agricultural activities. Of the 43,193 holdings that existed in the Basque Country in 1989, 5,465 specialized in dairy farming. This means 12.6 percent of the existing farms, and they are organized mostly around the work realized by the members of the family. The number of farms without paid workers rises to 5,317 (97.3 percent of existing milk producing farms). This group of farms is the target of this research. It is a homogeneous group as far as the kind of work involved, geographical location, and productive specialization.

The structural characteristics of dairy farms can be examined more precisely through data given by the Agrarian Census of 1989 (EUSTAT, 1991).[4] Some of the most relevant data are the following:

- The use of land is very much oriented towards feeding cattle: 90 percent of Utilized Agricultural Area (UAA) is occupied by meadows and pastures.

- An important number of farms (58 percent) has a small forest area: a 4.9 ha average of pine trees or other commercial timber.
- They have a very reduced area: the average per farm is 9.2 ha of Total Area and 5.5 ha of UAA.
- These are farms with a very reduced productive capacity: 70 percent of those with between 1 and 5 ha of UAA did not reach 4 European Size Units.
- The main form of land tenure was that of ownership: 68 percent of UAA is worked by proprietors.
- The number of people employed on dairy farms reached some 13,000; thus an average of 2.4 people work on each farm (compared to 1.8 for the whole sector).
- Some of the farmers have a low level of commitment. The 2.4 persons engaged on the farm do work that is equivalent to 1.4 Annual Work Unit (AWU). The reasons for this low amount of work are varied. In the case of the owners of the farm (who work an average of 0.68 AWU), this is due to their advanced age (27 percent of the owners are over 65), or to the fact that their main activity is outside the farm (20 percent). Thus, only 54 percent of the owners are younger than 65 and work exclusively on the farm.

This panorama refers to the year 1998, and dairy farms today may have very different characteristics since the Milk Quota Scheme has reduced the number of farms to a third of those existing in 1998. The farms that are least dependent on dairy production, those smallest in size, are the group that has been reduced in the greatest proportion.

An important aspect for an understanding of the external context, where the dairy farms are located, refers to how the milk is commercialized. Milk produced in the Basque region had the following destinations in 1999: 5 percent was consumed in the operation, 90 percent was delivered to dairies, and 5 percent was sold directly to the consumer (EUSTAT, 2001, p. 277). Although there are several dairies, there is a situation of quasi-monopoly because one of them, Iparlat, gathers around 82 percent of the milk delivered to the dairies. It is a firm owned by institutions, savings banks and farmers. Iparlat affects dairy farming through two mechanisms: the payment formula for the milk, and the management of the milk quotas. The payment formula is characterized by its having a low *starting price* and significant premiums and discounts according to the quality and quantity of the milk. Through premiums the starting price can be increased by up to 20 percent, and with the discounts, it can be reduced by as much as 23 percent. Since the final price obtained for the milk depends to a great extent on the quality and the quantity demanded by Iparlat, this system of payment results in two main consequences: it consolidates the most modernized farms, and it encourages small farmers to abandon milk production since it is no longer profitable. The second mechanism is through the management of the milk quotas. Iparlat is able to collect and reallocate the surplus milk from farms, which exceed their allocated quotas thanks to those farms that do not reach their milk quotas.

A final decisive element for understanding the link between the dairy farm and the external context is the agrarian policy implemented by the Basque institutions. Perhaps the two most relevant measures are the implementation of the Milk Quota Scheme, and the Investment Aid programme. The Basque Government is encouraging small farms to join the schemes for abandoning milk production. The quotas recovered through these schemes are then distributed freely among farmers who have carried out reforms on their farms, or who are producing more milk than that allowed by the quota system. In other words, this policy favours the enlargement of some farms. The second relevant agrarian policy is the Investment Aid Scheme. Through this measure, the provincial authorities are giving considerable amounts of aid for modernization to the more viable farms. These grants are directed towards genetic improvement, mechanization and the construction of cowsheds.

We may conclude from previous analysis that dairy farms in the Basque Country tend to have a small productive dimension because existing non-farming incomes allow farmers to reproduce the farm and the family. Agrarian policies and Iparlat try to eliminate the smallest farms from production through different mechanisms. The question here is to know how family farms react to this institutional power and, more precisely, whether different innovations in production, at the farm level, depend on the family strategy.

Family Strategies and Productive Innovations

Since the family typology reflects different family strategies, in this chapter a brief description of productive changes introduced by each type of family will be presented.

Aged Family

It is a family whose *central person* is over 65 or is between 50 and 64, and its incomes come from a pension, mainly from an early retirement scheme. This family has no heir. In 53 percent of the cases there are no children, and in 38 percent there is an offspring over 21 working only a few hours a day. This type of family has a rather small farm. An average of 4 hectares of grassland, 6.2 Cattle Units (it is an index where each type of animal has a different weight), and the total Horse Power of their machinery used in the meadows is only 11 Hp. Its small dimension can be explained by the scarce work available. More than half (56 percent) of these farms lack a tank for keeping the milk refrigerated and hygienic. The scarce volume of production and the investment required for buying and installing such a tank (fitting electricity, preparing the premises for its installation, etc.) does not make it a profitable acquisition. Nonetheless, since 68 percent of milk produced by these families is sold through dairies, the absence of the cooling tank indicates that they will obtain a great reduction in the milk price due to the lack of bacteriological quality. The price per litre obtained would be much less than that obtained by other farmers. Fortunately, this type of family does not depend only on an agrarian income. The pensions they receive are almost as important as their agrarian incomes, representing some 41 percent of their total income. In this way, pensions are contributing to the renewal of personal and productive consumption on these farms. Because of the non-existence of a successor, productive changes introduced by these families attempt to maintain the family income through some of the following ways: increasing their agrarian income by the direct sale of milk; reducing personal and productive expenses; and joining a Scheme for Abandoning Milk Production.

Adult Family, Exclusive Commitment, with Successor

It is made up of families where the *central person* is between 50 and 64 years of age and works exclusively on the farm together with an offspring over 21 dedicated exclusively to the work on the farm. It is the kind of family with the highest working capacity. In these families, the reproduction of the new family will be through the farm. The great dependency on agrarian income and the high availability of labour may explain why this type of family possesses a big farm (10.1 ha, 22 CUs, and 48 Hp in machinery). This suggests that these are families that have invested a lot and would find themselves deeply in debt despite benefiting from public funding for the modernization of the farm. Farms with this type of families are not only bigger, but also reach higher levels of intensification of production. On average there are some 3 CUs and 6.6 Hp per hectare. Productive changes are oriented towards the increase of its productive dimension: increase in production (77 percent of these farms have increased their Cattle Units between 1980 and 1985), and the improvement of technology (85 percent had a milk cooling tank in 1985). In other words, these families tend to modernize the farm. Their ideal type of farming is that of a highly modernized and specialized farm. They do not want other models of farming based on a diversification of incomes or on-farm transformation: 'I'm not convinced by rural tourism, or by selling 50 or 100 cheeses per month. In my case such income covers superfluous expenditure, but it doesn't solve any problems. I think that nowadays, those of us who produce milk, live from the milk ... yes, yes. We are dedicated to nothing else' (2:16).

Adult Family, Exclusive Commitment, without Successor

This type of family only differs from the previous one in that there is no successor. The farms belonging to these families show some characteristics similar to the previous type, but they are smaller in terms of hectares, CUs, and Hp. It seems that the presence of a successor is related to the size of the farm. These appear to be well-prepared farms which do not grow more due to the lack of an heir and which cannot reduce their productive capacity because they depend on their agrarian income for covering their costs. These families are 'stagnant.' They do not modernize, as do the previous ones,

Table 4.2 Productive characteristics of dairy family farms by type of family. Average
values in 1985

	Aged family	Adult family, exclusive, successor	Adult family, exclusive, without successor	Adult family, partial, without successor	Young family, exclusive	Young family, partial
Hectares of UAA	4.0	10.1	7.8	4.8	11.0	5.9
Cattle Units	6.2	22.0	16.8	8.7	24.4	9.1
Horsepower	11.0	48.0	33.0	26.0	45.0	28.0
Cattle Units/Ha	2.9	3.0	2.8	3.0	3.3	1.7
Horsepower/Ha	4.0	6.6	4.7	9.2	5.9	6.1
Sources of income:						
% from agriculture	57.0	92.0	95.0	44.0	94.0	39.0
% from pensions	41.0	7.0	4.0	1.0	5.0	6.0
% off-farm work	3.0	-	1.0	55.0	1.0	55.0
% delivered to dairies	68,0	63,0	66,0	69,0	59,0	52,0

given the lack of a successor and the old age of the parents: 'Yes, if there were younger
people, yes. Some years ago I was going to make a pavilion, but there was a nephew
at home who didn't like cows, and later he went to work in Bilbao. I said ... pavilions
now ... at 56 years old, am I going to make investments? Things are running down, in
the past we had a lot of cattle and now considerably less. And increasingly less ...' (3:33).

Adult Family, Partial Commitment, without Successor

This family has a reduced size of farm. Its productive capacity, however, is slightly
higher than the *aged family*. Though this type of family does not have much machinery in absolute terms, its reduced area makes it the type of family with the highest
number of Hp per hectare (9.2 Hp). It seems that these families try to compensate for
the lack of labour with greater mechanization. Capital investment in these farms is
high and since they receive little or no public economic support, due to the fact that
they are part-time farmers, we must conclude that income generated by off-farm employment is invested in the farm. Since this income is contributing to the renewal of
personal and productive consumption, it seems that this domestic group is interested
in maintaining the farm's activity in the short-medium term. Although agrarian income is reduced, we are dealing with the type of farm which least depends on this,
since 55 percent of its income comes from off-farm work. Lastly, this is the type of
farm which most tends to sell milk to the dairies (69 percent of milk produced). The
intensity of work that is required for direct selling and the absence of a successor are,

Table 4.3 Introduction of productive changes by type of family between the years 1980 and 1985

	Aged family	Adult family, exclusive, successor	Adult family, exclusive, without successor	Adult family, partial, without successor	Young family, exclusive	Young family, partial
Increase in Cattle Units	36	77	48	40	76	59
Increase in Horse Power	12	52	20	25	42	24
Decrease % milk to dairies	9	9	6	10	13	6
Acquisition of milk cooling tank	44	85	73	55	90	47

Note: Association between each productive change and the type of family has been measured through the Chi-square test. The association is significant at a 95 percent level of confidence, except for the evolution of the percentage of milk sold to dairies (significance level is 0.70).

perhaps, the reasons that prevent them from having the manpower needed to carry out the direct sale of milk. On those holdings where the farmer combines work on the farm with an off-farm job, the couple is subjected to a hard rate of work. It is the mother who carries most of the workload and who makes it possible for the farm to exist. There are many reasons why these couples *sacrifice* their lives, but in many cases farming generates the additional income they need to provide their children with a higher level of education. The participants in the focus groups pointed out how important the mother's role is for this type of family. We should note that they did not refer to the women's role but to the mother's role: 'I recognize this and I say it ... the merit is not ours. I've said this many times. The merit belongs to our wives. The man is always held up as deserving credit ... The man hasn't done a thing. Without his wife he wouldn't have done a thing. If the man has one of those wives who says: you busy yourself with your cows and all that, then you're stuck on your own ...' (10:23).

Young Family, Exclusive Commitment

These families hold similar farms to the *adult family, exclusive commitment, with successor* type. They are large farms (here we find the farms with the biggest number of hectares and cus), they depend on farming for their living (94 percent of the total income comes from agrarian production), and have highly professionalized agrarian practices (90 percent have a milk cooling tank, and 78 percent practice cattle breeding). It is also the type of farm that has a more intensive production because the high price of the land makes it difficult to increase the size of the holding. The cattle load represents some 3.3 CUS per hectare. This greater intensification may be due to recent investment, which in turn forces them on to a greater production and productivity. These families carry out modernizing changes on the farm such as increasing the volume of production (76 percent have increased CUS) and technological improvement (90 percent have a milk cooling tank). These are the farms that have most reduced the percentage of their production sold to dairies (a reduction of 13 percent between 1980 and 1985). This productive change may be explained because their high economic debt forces them to increase income from the sale of milk.

Young Family, Partial Commitment

This constitutes a domestic group where the family's reproduction is achieved through an off-farm job. They keep the farm on after finding another occupation because they enjoy the work with cows. A vocational element for the continuation of the farm seems to be present in this type of family. In fact, this is the type where agrarian income contributes least to the family's income. This family shows certain characteristics, which make it different from the rest. Firstly, the most extensive form of farming (only 1,7 CUS per hectare). Secondly, it has a much more diversified production, where calf fattening is of some importance (on 76 percent of the farms). Finally, this is the type of family which most tends to sell its milk directly to consumers (only 52 percent of the produced milk is sold to the dairies). The above suggests that this family sells directly more frequently because they need to compensate for the lesser income from fattening by increasing their earnings from the sale of milk.

Conclusions

Productive changes introduced by family dairy farms in the Basque Country seem to depend on a number of conditions, such as the farm's capacity for generating an adequate level of income, the possibility of finding an off-farm job, the family coming together on a common plan or strategy, whether the would-be successor sees the working conditions inherent in dairy farming in a positive way, or the mother's role. It has become clear that the role played by mothers over 49 years old is crucial, because of their self-sacrifice both in supporting the offspring's decision to take over the farm, and in maintaining productive activity where the farmer only works part-time. These mothers, rather than women in general, seem to be a key aspect in family farming.

Families take decisions evaluating their own resources (labour, economic, desires, etc) and the external context within which they are located (prices, policies, possibilities of an external job, etc). After considering all these elements, the domestic group decides whether a successor will continue farming. The concept *family strategy* summarizes possible decisions, and the typology of families is a good indicator of the possible strategies. Since the holding is an economic resource serving family interests, farm changes will vary depending on the family's decision. Therefore, although family farming is a homogenous type of farming, family farmers are not a homogenous type of family.

The type of family is a good indicator of this reproductive plan, and forms an appropriate criterion for understanding the present characteristics of farms and future productive innovations. The Basque public institutions must be aware of the differential behaviour of family farming facing the agrarian policies they draw up, and they need to consider the domestic group rather than the farm as a unit of analysis. The functioning and changes of the Basque farms cannot be understood from a strictly productive and economic logic.

Although the concept of family strategies seems to be adequate, further thought must be devoted to understanding aspects such as the decision-making process in the domestic group, the type of strategies available to family farming (productive, personal, collective etc.), or the influence of the holding's location on the strategies available.

Finally, we must conclude that the *agrarian question* in the Basque Country, the development of capitalism in agriculture, is not leading to the consolidation of capitalist farming (that with salaried workers), but to the differentiation of family farming according to its degree of modernization. This differentiation process is not spontaneous, but rather a consequence of the model of agricultural development being encouraged by the Basque institutions, through measures such as the re-allocation of production rights and the Investment Aid scheme.

Notes

1. Additional information on this research approach may be found in the author's Ph.D. thesis (Mauleón, 1998).
2. This approach is increasingly popular for understanding agriculture since most firms are structured on a family basis. There is a growing tendency towards its application for understanding the productive changes in family farming in the European context (Martini and Pieroni, 1987; Blekesaune, 1991), in Latin America (Chonchol, 1990), or the United States (Schulman, 1994).
3. It is important to note that a seventh type of family is not included in the typology: *Adult family, partial commitment, with successor.* There were only two cases in the survey, and they were eliminated from the analysis. It seems that on those farms where the father works on a part-time basis, succession is not common.
4. The data available from the latest Agrarian Census, of 1999, is still partial and provisional.

References

Blekesaune, A. (1991), Changes in ways of making a living among Norwegian farmers (1975–1990). *Sociologia Ruralis* 31 (1) pp. 48–57.

Bouquet, M. (1982), Production and reproduction of family farms in South-West England. *Sociologia Ruralis* 22 (3/4), pp. 227–244.

Cardesín, J. Mª. (1992), *Tierra, trabajo y reproducción social en una aldea gallega (s. XVIII–XX); muerte de unos, vida de otros.* Madrid: Ministerio de Agricultura, Pesca y Alimentación.

Chayanov, A.V. (1985), *La organización de la unidad económica campesina*. Buenos Aires: Ediciones Nueva Visión.

Chonchol, J. (1990), Modernización agrícola y estrategias campesinas de América Latina. *Revista Internacional de Ciencias Sociales* 124, pp. 143–160.

Djurfeldt, G. and Waldenstrom, C. (1996), Towards a theoretically grounded typology of farms: a Swedish case. *Acta Sociologica*,39, pp. 187–210.

EUSTAT (1991), *Censo Agrario de la Comunidad Autónoma de Euskadi. 1989.* 4 volumes. Vitoria: Instituto Vasco de Estadística.

EUSTAT (2001), *Anuario estadístico vasco, 2001.* Vitoria: Instituto Vasco de Estadística.

Evans, N. and Ilbery, B. (1996), Exploring the influence of farm-based pluriactivity on gender relations in capitalist agriculture. *Sociologia Ruralis* 36 (1), pp. 74–92.

Galeski, B. (1977), *Sociología del campesinado*. Barcelona: Ediciones Península.

García Ramón, M.D. (1990), La división sexual del trabajo y el enfoque de género en el estudio de la agricultura de los países desarrollados. *Agricultura y Sociedad* 55, pp. 251–277.

Goodman, D. and Redclift, M. (1985), Capitalism, petty commodity production and the farm enterprise. *Sociologia Ruralis* 25 (3/4), pp. 231–247.

Greenwood, D.J. (1976), *Unrewarding wealth. The commercialization and collapse of agriculture in a Spanish Basque town.* Cambridge: Cambridge University Press.

Friedmann, H. (1978), World market, state and family farm: social bases of household production in the era of the wage labor. *Comparative Studies in Society and History* 20, pp. 545–586.

Friedmann, H. (1980), Household production and the national economy: concepts for the analysis of agrarian formations. *The Journal of Peasant Studies* 7, pp. 158–184.

Marini, M. and Pieroni, O. (1987), Relación entre la familia y el entorno social. Tipología de las familias agrícolas en una zona marginal (Calabria). Arkleton *Cambio rural en Europa*. Madrid: Ministerio de Agricultura, Pesca y Alimentación, pp. 205–244.

Marsden, T. et al. (1989), Strategies for coping in capitalist agriculture: an examination of the responses of farm families in British agriculture. *Geoforum* 20, pp. 1–14.

Mauleón, J.R. (1997), Survival strategies of dairy family farms in the Basque Country. Paper submitted for the XVII *Congress of the European Society for Rural Sociology*, Chania (Greece).

Mauleón, J.R. (1998), *Estrategias familiares y cambios productivos del caserío vasco.* Vitoria: Departamento de Presidencia del Gobierno Vasco.

Nalson, J.S. (1968), *Mobility of farm families*. Manchester: Manchester University Press.

Nooij, A.T.J. and Somers, B.M. (1986), Part-time farming in the Netherlands. Agricultural policy and the marginalization of a 'non-professional' farming group. Paper submitted at the XIII *Congress of the European Society for Rural Sociology*. Braga (Portugal).

Ploeg, J.D. van der (1985), Patterns of farming logic, structuration of labour and impact of externalization. Changing dairy farming in northern Italy. *Sociologia Ruralis* 25, pp. 5–25.

Potter, C. and Lobley, M. (1992), Ageing and succession of family farms. *Sociologia Ruralis* 32, pp. 317–334.

Schulman, M.D. et al. (1994), Survival in agriculture: linking macro- and micro-level analyses. *Sociologia Ruralis* 34, pp. 229–251.

Whatmore, S.J. et al. (1986), Internal and external relations in the transformation of the farm family. *Sociologia Ruralis* 26, pp. 396–8.

Whatmore, S. et al. (1987), Towards a typology of farm businesses in contemporary British agriculture. *Sociologia Ruralis* 27, pp. 21–37.

Chapter 5

Innovation Formation and the Practice of New Rural Partnerships in Finland

Torsti Hyyryläinen

The Budapest Declaration on Rural Innovation (2002) adopts a very critical approach to the concept of innovation as a component of rural development. The declaration believes in building up innovative activities, which stem from actors within civil society. The crucial way to facilitate and promote rural development is through new state–community relationships, which can both empower and discipline the efforts of local voluntary actors and groups. The purpose of this chapter is to continue this critical evaluation of the concept of innovation by concentrating on the question of how rural development issues can assume the form of problems that are relevant to rural inhabitants themselves. The accent will thus be on *problem formation*, and thereby also on *innovation formation*. I shall attempt by quoting Finnish examples to demonstrate the importance of such an evaluation for practical rural policy, and examine in particular new forms of co-operation, set up at different spatial levels within rural policy, referred to recently as 'new partnerships'. It will become evident at the same time that the question of innovation formation is closely linked to existing *power relations*.

The point of departure in this discussion is that the pronounced sectorization of our societies is a crucial structural factor affecting the formation of problems and innovations. This sectorization has already proceeded far too far and has placed further difficulties in the path of sustainable rural development. A solution to this problem has been sought through increased emphasis *on integrated rural development*, new forms of co-operation and *networking* that cross the boundaries between private and public organizations, and new forms of governance. A trend in the same direction is to be perceived in those representations, texts and speeches that lay emphasis on the significance of *new partnerships*. I am inclined to interpret the discussion of partnerships in this connection as attempts at practical application of the ideas aroused by new concepts such as *governance* and *social capital*. This would imply that the concept of partnership as such is by nature both a practical and political instrument for change (Stoker, 1997; Lowe et al., 1998; Goodwin, 1998; Shucksmith, 1998; Westholm et al., 1999; Geddes and Bennington, 2001; Grippen et. al., 2001).

How to Find Relevant Problems – The General Framework

Rural development is targeted at problems of various kinds, depending on how they are defined in different problem-solving situations. This issue can be approached via a general framework for rural innovation formation, the various aspects of which are all united by the question of *participation*. Understood in a broad sense, this is a matter of how well democracy functions and how rural inhabitants can be involved in decisions that concern them. Another essential aspect is the question of *place*, since places differ in the conditions that they provide for development. At the center of the framework lies the *creative question* 'what?', representing simultaneously the object of problem solving and the object of development activities.

Figure 5.1 A general framework for rural innovation formation

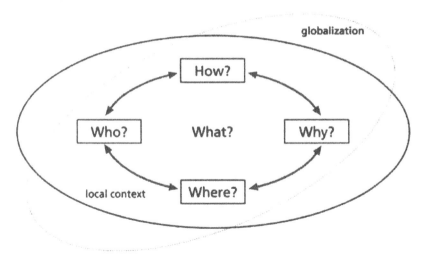

The fundamental question is how to find the relevant problems in a local context. When we speak of a local context in rural development we usually mean action carried out under local conditions by local inhabitants, the authorities, politicians and entrepreneurs. The problems encountered are consequently always quite specific ones and have to be solved in the context of practical situations, employing the best available knowledge and resources. To understand new innovations as a part of the problem-solving process, it is relevant to ask:

1. *who* is committed to investing human and other capital in the joint problem-solving process?
2. what are the motives, interests, goals and knowledge base of each participant? (*why?*)

3. *how* is participation to be organized and how extensively will it attract people to participate?
4. under what conditions are the practical problem situations formed, interpreted and assigned priorities? (*where?*)

Local conditions themselves, however, do not develop in isolation from wider changes taking place in the community and internationally. Globalization is having vast effects on rural areas, and local development environments are also being affected in numerous ways by the social, employment, taxation and rural policies adopted by national governments. Local rural developers are thus obliged to operate in situations created by forces that they are unable to influence.

The difficulties facing local development are extremely complex, in the nature of 'wicked problems,' which have no single definitive formulation but are unique to particular situations. This means in turn that the processes of solving these problems will also be complex and will often entail conflicts of interests. Some solution has to be found in each practical situation, however, mostly a compromise solution that 'everyone can live with'. Local developers are therefore faced with a challenging task from the very outset.

Sectorization and the Formation of Rural Problems

In Finland, as in many other countries, functions within society are organized primarily on the basis of specialized institutions, and even the public sector is professionally highly sectorized. In other words, each significant function within society, such as health care, employment, economic activities or education, is looked after by its own sector of the administration, which safeguards its own interests and develops its own activities in an independent and self-centered manner. In a sectoral society of this kind the principal paths of communication operate in a vertical direction – usually from the top down. Each sector is responsible only for its own results, strategies and development programmes regardless of the whole.

Sectorization is one of the fundamental characteristics of public administration in our modern social systems, and is as such essential for the administration of the exceptionally complex functions of large human communities, but it also gives rise to problems of its own. The most serious of these is that an excessively developed sectoral administration is apt to lose its grip on society as a whole, so that in the worst case no instance feels any responsibility for the overall outcome. The sectorization of the public administration can also be problematic for rural areas, as this administration usually plays a significant role in matters of rural policy.

Specialized organizations of this kind nevertheless have a habit of strengthening their own position as they achieve growth, and in their drive for greater efficiency they are apt to decline in their ability to: (a) look for alternatives, and (b) deal with individual problems and things that differ from the ordinary. This sector model is best adapted for dealing with routine matters and problems that are restricted to one branch of expertise. Sectors of this kind work well in 'mass production' situations, i.e., when the answers to the problems are already known, as it is then merely a matter of implementing the known answers on a sufficiently broad scale.

Our current social problems, such as rural unemployment, have nevertheless become highly complex, so that the expertise of specialist organizations is no longer enough. It is impossible to find a 'mass-produced' solution to unemployment, because the problems differ radically from one branch of the economy to another and from one geographical area to another.

Finland is a very sparsely populated country with great differences between its regions. There are only a few large cities apart from the Helsinki conurbation, but there are innumerable small country towns and built-up areas acting as local centers for rural areas. There are also several kinds of rural area, but they all have one feature in common: they have substantially fewer potential actors (inhabitants, enterprises or institutions) than the urban areas. This in turn means that the culture of corporate action is different, being more closely knit in many ways.

With this in mind, it may be said that the crucial point of departure for all rural development is the pooling of resources through 'cross-border' co-operation and interaction. One has to look hard to find these rural development resources in the first place, and once they are found, it is important to gather together a 'critical mass' by ranging over a wide area and a broad spectrum of organizations. The scarcity of resources means that every avenue has to be explored and much hard work has to be done to earn every grain of success.

The same scarcity of intellectual and social resources also implies that it is impossible to follow a sectorized pattern in rural development. Instead, activities have to be based on a search for synergy, for a comprehensive inter-sectoral commitment to development. In the sparsely populated Finnish context it is abundantly clear that without an integrated approach all efforts will remain ineffective and impacts inconsequential.

It may be claimed that the sectorization of societies has gone too far and that partnership is needed to counterbalance it. *Partnership* can be seen as a mode of action that employs a broader *knowledge base* and takes note of particular local features. According to Jane Nelson and Simon Zadek, new partnerships can be defined as: 'people and organizations from some combination of public, business and civil constituencies who engage in voluntary, mutually beneficial, innovative relationships to address common societal aims through combining their resources and competencies' (Grippen et al., 2001, p. 8). In many case rural partnerships are expected to be: consensus building; promoting the building of local strategies; facilitating co-ordination actions; giving access to different skills; promoting innovation; strengthening local identity and competitiveness (Westholm, 1999, p. 14).

New structures are required that operate between sectors, and co-operation and development procedures are called for. All this presupposes knowledge and expertise that crosses sectoral boundaries. It is also becoming essential to employ new concepts and to create new meanings for co-operation at the national, regional and local levels.

Towards an Integrated Rural Policy

In the sphere of Finnish rural policy a search has been going on for a long time for a solution to the problem of creating a truly integrative policy. From experience we know that this is a matter of achieving both structural and functional changes at various levels. It is crucial to point out the need for a synergy of *national, regional* and *local* level activities in order to create adequate conditions for successful rural living (Hyyryläinen and Uusitalo, 2002).

As far as national policy is concerned, it is essential to increase interaction between the ministries involved with rural affairs and to clarify the relative powers and responsibilities of central government and the regional authorities. In Finland some progress has been made at the central government level. Rural policy is basically an integrated policy, and has been co-ordinated at the national level since 1992 by a Rural Policy Committee containing representatives from a number of ministries that has been set up in order to draft rural policy and plans for its implementation. This committee has a number of working groups concerned with particular aspects of rural policy, and produces its own publications. It also funds rural research and development projects, draws up rural policy programmes and supports discussion on rural policy in society at large. This represents a unique form of long-term consultation between ministries, and has already proved invaluable.

Every municipality in Finland is responsible for the development and welfare of its own area and residents. Both the state and the joint authorities representing the municipalities (i.e., the Regional Councils) are responsible for regional policy as a whole. Although the *Regional Councils* have the statutory duty of promoting regional development, they have never been given adequate resources for performing this duty. For political reasons, the majority of the finance is channeled via another system, the *Employment and Economic Development Centers,* which also operate on a regional basis. This means that the division of power and responsibility in the regions is by no means ideal for rural policy or for the preparation and implementation of regional development programmes.

It is important to note that decision-making processes at the regional level differ between these two organizations. Those taking the decisions in the Regional Councils are representatives of individual municipalities, frequently officials or prominent political figures, whereas the Employment and Economic Development Centers are regional organizations of the state in which decisions are taken by groups of civil servants. Upon the introduction of EU programme-based policy, efforts were made to remedy this situation by creating *new regional partnership groups,* involving in each case representatives of the region, its local authorities, the state authorities, people from business and industry, the labour market organizations and other interested parties, and allowing for its meetings to be attended by representatives of the Ministry of the Interior and the European Commission. These partnership groups take decisions on projects connected with more extensive EU programmes. As far as the local inhabitants are concerned, however, this has remained a distant and unfamiliar process, which allows scarcely any opportunity for participation.

At the local government level, Finland is divided into about 448 units known as municipalities. It has become increasingly obvious in recent years, however, that the smaller rural municipalities are unable to muster the resources necessary for development, and they have begun to form groups for this purpose. This has led to the emergence of somewhat larger functional units, known as *sub-regions*, as a significant new local level for regional development. There had certainly been an opening for units of this kind below the level of the region, not least as areas of suitable size for the operation of local action groups (LAGS).

It would thus seem that regional development in Finland during the present decade will be a matter of co-operation between local authorities at the sub-region level and measures planned and undertaken by local action groups, for it is precisely within these groups that the inhabitants, communities, enterprises and local authorities will have the best opportunities for crossing sectoral boundaries and creating common strategies. The crucial phenomenon is that the municipal authorities have taken new Local Action Groups as forums of local development processes. This has called for high levels of networking, interaction and communication skills, for the nucleus of such activity lies in the joint formulation of strategies and the shaping of local initiatives into projects (Hyyryläinen and Rannikko, 2000; LEADER II evaluation, 2002).

New Partnerships – New Innovations?

In the light of the above it can be said that significant new forums for partnership were constructed at the central government, regional administration and local levels within the Finnish rural policy system in the course of the 1990s, which possess a potential for generating innovations. In summary, it may be noted, however, that the new partnerships implemented at the regional level have entailed far less innovative elements than those at the central government or local level, as the regional forums for partnership set up in order to administer EU programmes have so far been mainly forums for local elites, and scarcely any new participants have been obtained for these bodies dominated by civil servants and politicians. Since participation is based on hierarchical position in certain regional organizations, the members largely represent the same motives and interests as earlier and their ways of working are highly conservative and dominated by established routines. It is difficult to gain information on the activities of these groups, and thus they have remained an unknown quality as far as the general public is concerned. In practice, however, it is these local elites that distribute the vast majority of the EU development projects funds that are available in Finland.

The national Rural Policy Committee set up in 1992 is a significant exception in the Finnish culture of political decision making, in that it has not only constructed a link between various ministries but it has also made a determined effort to extend its network in the direction of both experts and local rural actors. Considerable devolution has taken place in the processes by which decisions are prepared, and the committee can be regarded as a notable pillar of rural policy discourse in Finland at the present time. It can thus be said that rural innovations in this country are worked up at both the central administration and local level through the creative processes of constructive new partnerships.

Partnerships are needed to counterbalance our strongly sectorized society. New structures are required that operate between sectors, and co-operation and development procedures are called for. All this presupposes knowledge and expertise that crosses sectoral boundaries. But it is also a question of policy and political strategies. For example, a partnership project was taking place in the employment sector in 1996–1999 at the same time as in the field of rural policy (Ministry of Employment; see Luostarinen and Hyyryläinen, 2000). If we compare the practical details of the implementation of rural partnerships with the situation prevailing in the national employment policy and its programmes, the following differences can be seen:

- The *concept* of partnership was implemented in a much more purposeful, systematic and diversified manner in the field of rural policy, and it was
- *Promoted* more actively by means of a number of development projects.
- *Local* action groups were expected to come up with clear strategies and concrete implementation schemes of their own.
- It was not believed that partnership could be implemented in the field of rural policy without the creation of *new structures* at the local level, and thus each local action group was expected to organize itself into a legally constituted association or co-operative. And finally, the promotion of partnership has been a long-term endeavour in rural policy, even though it has advanced in small steps (see Hyyryläinen, 2000).

Comparison between these policy sectors (employment policy and rural policy) also demonstrates that partnership cannot be brought about by local actions alone but calls for broader political guidelines and well-planned supporting measures. Although the implementation of partnership in the field of rural policy cannot be explained entirely by the fact that it does not have to suffer so much from any of the disadvantages of management by results applied to a single sector of the administration, something must be gained by being able to concentrate on local development as a whole.

Finnish rural policy has made significant practical efforts to create new local structures and partnerships. Perhaps the most important single factor in this has been the setting up of a nationwide network of Local Action Groups, partly under the stimulus of the European LEADER programmes (LEADER II and LEADER+) and the equivalent national POMO programmes (POMO and POMO+) since 1996. At present there are 58–60 of these action groups in different parts of the country, and experiences of their work have been encouraging, in that more extensive co-operation is now taking place between the local inhabitants, communities, entrepreneurs and municipal authorities and new people and ideas have been mobilized. These new local partnerships have blended in well with the long tradition of local (municipal) government in Finland and contribute to the continued strengthening of local participation.

- 556,000 participants in activation and other meetings.
- 480 new enterprises created.
- 3,900 new jobs created, of which over 700 were full-time jobs.
- The highest proportion of projects implemented (27%) were for improving the environment or living conditions.
- Artisan, service and small enterprise projects amounted to 20% and tourism projects to 19.7%.
- There were only fourteen trans-national projects.
- Two-thirds of the projects were carried out by communities or companies.
- Among public organizations, the municipal authorities were responsible for the largest number of projects, 257.
- The work of the local action groups was a new thing for everyone at first and required much learning.
- The local emphasis was visible in the goals set for the local strategies and programmes.
- The numbers of participants increased as the work progressed.
- The steering committees of the local action groups felt that they were working independently.
- Ways of working became more flexible with time and adjustments could be made for the bureaucracy involved.
- Contacts between local actors and the authorities, especially sources of finance, increased with time.
- Co-operation with the local authorities strengthened and confidence in their actions increased.
- New people became involved in the local development work, but even so,
- The broader-scale project work was left to an excessively small number of participants.
- Co-operation between the local authorities and various associations increased.
- The work of the local action groups complemented the range of economic development measures available to the local authorities.
- The state's regional administrators became more favourably disposed towards these projects as the process advanced.
- The goals for the development of entrepreneurship were not achieved in all respects, but
- Co-operation and networking between enterprises did improve.

A fundamental ex-post evaluation of LEADER II activities in Finland carried out in 2000–2001 indicated that a total of 22 Local Action Groups had been responsible for implementing 3050 projects between 1996 and 1999. The geographical areas served by these groups had a combined population of 814,000 persons living in 157 municipalities, with each group covering 2–16 municipalities. The mean population of the area covered by one group was 37,000 inhabitants (LEADER II evaluation, 2002).

Total expenditure on LEADER programmes was 78.32 million Euros, of which 73% was public funding from EU or national sources. The national evaluation of LEADER II programmes in Finland documented the impacts achieved as follows.

The evaluation concluded that LEADER II marked an important investment in local *social capital* in Finnish rural areas. The principal observation as far as the present topic of innovation formation is concerned was that LEADER II was very successful in activating local participants by comparison with other forms of development, and it is particularly notable that no other developmental approach has been able to achieve comparable results at the local level (LEADER II evaluation, 2002, pp. 83–84).

Local and Personal Commitments in Scope

Local activity, personal commitments and investments in *social capital* are of great importance in the context of rural policy. Both physical and human capital is to be found in the countryside, but both are sparsely distributed. Given the correct procedures, these scarce resources can be identified more efficiently and gathered together so that they can have a creative influence on each other.

Many Finnish villages have had voluntary Village Committees to attend to their development needs since the 1970s (at present 3,200 in number), and the structure of this traditional activity is now being revised. The rise of project work carried out on a professional, paid basis alongside the ancient tradition of unpaid *talkoo* work that has prevailed in the Finnish countryside has greatly altered internal relations and responsibilities within these village activities, and there is a general awareness in the villages of the nature of the challenges facing these activities and of the need for new people, ideas and development tools.

Village activities are seeking a new impetus from expanded co-operation. Neighbouring villages, experts of various kinds, local action groups and even international actors are proving to be significant partners for village activists and entrepreneurs nowadays. The co-operation networks are still relatively weak, however, and contacts between actors are somewhat haphazard, so that the new forms of co-operation may be looked on as attempts to solve this problem. One urgent question that comes to mind is that of how local initiative can be linked in a flexible manner to a sub-regional development environment.

One important precondition for co-operation at the sub-regional level is that the sub-region should possess the corporate networks and responsibility structures on which co-operation can be built up. *Social capital* implies a capacity for working together with others in groups, organizations and social networks for the common good, and can be learned only by experience of such interaction and activated by establishing interaction and networking. The promotion, maintenance and utilization of social capital calls for skill and hard work, together with the correct methods and operative models.

It is important to note that both the structural and individual conditions required for the emergence of this social capital already exist. The process calls for norms and trust structures and also for personal commitments and personal investments. The crucial question is how these two sets of conditions can be met simultaneously.

The point of departure to be considered below is that the villages are still of importance as *places* for living and conducting business and that they have a special meaning for the people who live in them. They are not the same social communities as they used to be, of course, not even in Finland, for people nowadays tend to have numerous varying identities, only some of which are associated with their own immediate living environment. But even so, the villages can be looked on as nodes at which social capital can be generated through thinking, imagining and doing in connection with *common issues and innovations*. The essential thing as far as the local construction of social capital is concerned is to achieve a genuinely new, creative problem-solving process.

It is crucial to develop models for analysing interaction between local actors in a more comprehensive manner. This would enable a picture to be obtained of the principal actors and modes of operation involved in successful local development and of the main forms of expertise and support that villagers require for their own development projects.

Figure 5.2 A network model of local partnership for rural innovations

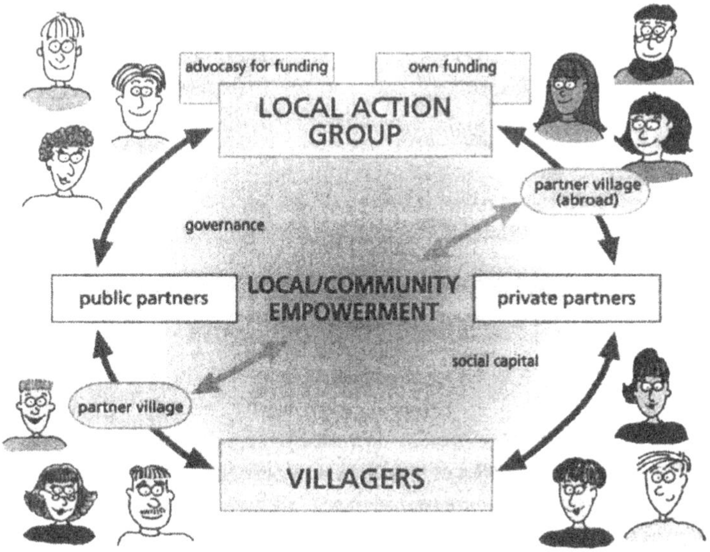

The above model places emphasis on the *villagers'* perspective, i.e., that of the people who live and work in the village permanently or on a part-time basis. A sustainable point of departure is provided by a set of aims decided on communally by the villagers, the resources available and the degree to which these resources are recognized. The villagers' investments in the network take the form of their own commitment, human capital, local knowledge and voluntary *talkoo* work that they put in on the projects.

Processing of the villagers' initiatives and implementation of the resulting projects calls for co-operation between *public* and *private partners*. The purpose is not to set out to develop and implement ideas alone, but always through partnerships. These partners, who may be individual sympathizers or experts from various organizations, invest their own capital in the joint projects in the form of their knowledge and skills, either administrative or substantive.

The third instance is the *local action group*. These groups composed of inhabitants and entrepreneurs living or operating in the area in question and representatives of village committees, societies and local authorities (municipalities) form the broader development environment, a 'field of creative competition,' in which various ideas, transformed into projects, can be mutually compared. The local action group invests the (inter-sectoral) knowledge available to it through its actors in the network and channels public funds into development.

A local development network arises out of practical *commitments* to the vitality of the countryside and its villages and simultaneous *investments* in it by numerous private, public and community instances. The creation of modes of operation and mutually agreed rules and regulations is a matter of strengthening the *social capital* available locally. The mobilization of new people and new ideas is a major precondition for success, but this in turn requires strengthening of the responsibility structures and the provision of better operational models, for without these local development is likely to become too great a burden for the few people who are active in it.

Once this local network model can be made to work in practice, development will become a continuous process in which initiatives arising out of the needs, ideas and commitment of the actors in the villages are processed into projects together with partners and returned to the local action group for evaluation and assessment relative to the funding available. Those projects that gain acceptance in the action group as being in accordance with its strategy will be implemented in co-operation with the partners.

Conclusions

Rural policy is a broad field in which conditions for success need to be created in many ways and by many instances. It is necessary at the national level of regional policy to realize that regional differences place differing demands on practical development, whereupon a differentiated rural policy can be invoked to construct modes of operation that are sensitive to the characteristic features of a particular area. The Finnish rural policy shows in practice that the statements of the Budapest Declaration point in the right direction but are not easy to implement.

Innovative governance requires a clear shift in political programming and decision making, towards recognition and empowerment of local rural population groups, NGOs and other actors within civil society. We can learn from the Finnish case that this change cannot be implemented by EU or governmental agencies in a top-down manner only, but has to be approached through networking and the creation of social and political conditions which support independent initiatives, and by allowing these to both control and evaluate the development process independently as far as possible, and with less and less bureaucracy (Budapest Declaration 2002).

Local activity and personal commitments and investments in *social capital* are of great importance in the context of rural policy. According to a national evaluation of local action group activities in Finland it could be said that investments of the LEADER type were important in promoting local social capital in Finnish rural areas. Our research results suggest that the European LEADER programmes and the corresponding national POMO programmes have encouraged new people to engage in development work and have offered opportunities for putting innovative ideas into practice. Given the correct procedures, these new human resources can be identified and gathered together so that they can have a creative impact on local development.

Rural areas can provide a different development environment for citizens' initiatives, one that is of smaller proportions but more sensitive. At best, such an environment could abound with social innovations that respect the individual but at the same time take account of the nature of the community and its needs. It is particularly important to recognize the special character of rural areas and to allow local development structures and procedures to continue to evolve on the basis of existing experiences.

According to the Budapest Declaration, most discussions of innovation understand it either as the property of scientific and technological experts or the property of entrepreneurs acting in markets. Although these technical and economic forms of innovation are important, it is crucial to note that social and institutional forms are needed too. This is very clearly seen in the Finnish case, but at the same time it is an aspect that seems to be neglected in political discussions. The Declaration emphasizes in particular institutional innovations within civil society, which are the intended or unintended outcomes of efforts by social actors (rural households, voluntary organizations, social movements, etc.) to create new or 'alternative' organizational models, e.g., for the production or circulation of food, for the management of environmental resources, or for leadership or democratic empowerment of rural communities (Budapest Declaration, 2002).

One thing that is clearly visible in our analysis of innovation formation on the basis of the situation in Finland is that the definition of an innovation is very much a question of *who* is in an appropriate position to participate in the definition process. This means that innovations constitute a core area of political activity. These definition relationships vary greatly within the new partnerships that have been set up at different levels in the rural policy system. Partnership can be implemented in a manner that does not in effect imply any major changes in the prevailing power structures.

The participants in these genuinely new partnerships are committed to entering bravely into interaction with new partners, so that the circle of participants can be extended and new perspectives and ideas can be introduced into the problem-solving process. This means at the same time, of course, that the process also becomes more complex, and also more difficult, especially at the beginning. In the context of Finnish rural policy these genuine new partnership networks have been created most of all at the local level, although, perhaps somewhat surprisingly, they have also emerged within the central government. These social innovations appear to have caught on least well in the intermediate circles, those of the highly sectorized regional administration.

Finland's new, Europeanizing regional policy has had its most significant impact at the local level, where the municipalities traditionally play a crucial role. The investments made in the work of the local action groups have been more efficient in stimulating local activity than any other development measures undertaken in the 1990s. This activity can open up new opportunities for developing systems of local participation, enabling practical rural policies to be directed more obviously towards to problems which people have a genuine interest in solving.

References

Budapest Declaration (2002), *The Budapest Declaration on Rural Innovation*. Final Conference of COST Action 12. Budapest.

Geddes, M. and J. Bennington (eds) (2001), *Local Partnerships and Social Exclusion in the European Union - New Forms of Local Social Governance?* London: Routledge.

Goodwin, M. (1998), The governance of rural areas: Some emerging research issues and agendas. *Journal of Rural Studies* 14 (1), pp 5–12.

Grippen, C., K. Pinnington and A. Wilson (2001), *The role of Central Government in developing New Social Partnerships. The findings from seven European countries.* The Copenhagen Centre, Denmark.

Hyyryläinen, T. (2000), To find the right problem. Comments on the LEED Finnish study report and results of the Finnish Action Research. Paper 29.9.2000 presented at the Conference on Partnerships. 2–3. Oct. Helsinki. www.mtkk.helsinki.fi/henkilosto/hyyrylainen/torsti.htm

Hyyryläinen, T. and P. Rannikko (eds) (2000), *Eurooppalaistuva maaseutupolitiikka. Paikalliset toimintaryhmät maaseudun kehittäjinä.* Vastapaino. Tampere.

Hyyryläinen, T. and E. Uusitalo (2002), Recent rural restructuring and rural policy in Finland. In Halfacree, K., I. Kovach, and R. Woodward (eds) *Leadership and Local Power in European Rural Development.* Aldershot, Ashgate, pp. 230–254.

LEADER II evaluation (2002), *LEADER II - yhteisöaloiteohjelmien jälkiarviointi.* University of Helsinki Seinäjoki Institute for Rural Research and Training. Reports 82.

Lowe, P. et al. (1998), *Participation in rural development.* Centre for Rural Economy, Department of Agricultural Economics and Food Marketing. University of Newcastle upon Tyne, Research Report.

Luostarinen, S. and T. Hyyryläinen (2000), *Uudet kumppanuudet.* University of Helsinki Mikkeli Institute for Rural Research and Training. Publications 70.

Stoker, G. (1997), Public–private partnerships and urban governance. In G. Stoker (ed.), *Partners in Urban Governance: European and American Experience.* MacMillan, London. pp. 1–21.

Shucksmith, M. (1998), Rural and regional policy implementation – Issues arising from the Scottish experience. In Saukkonen, P. and H. Vihinen (eds) *Rural and Regional Development.* University of Helsinki, Mikkeli Institute for Rural Research and Training. Publications 61. pp. 13–41.

Westholm, E., M. Moseley and N. Stenlås (eds) (1999), *Local Partnerships and Rural Development in Europe. A Literature Review of Practice and Theory.* Dalarna Research Institute, Sweden, in association with the Countryside & Community Research Unit, Cheltenham and Gloucester College of Higher Education. Falun.

Westholm, E. (1999), Exploring the Role of Rural Partnerships. In Westholm, E, M. Moseley and N. Stenlås (eds), *Local Partnerships and Rural Development in Europe. A Literature Review of Practice and Theory.* Dalarna Research Institute, Sweden, in association with the Countryside and Community Research Unit, Cheltenham and Gloucester College of Higher Education. Falun. pp. 13–24.

Chapter 6

Power/Knowledge in the Discourse of Rural/Regional Policy

Hans Kjetil Lysgård

The main argument in this chapter is that rural/regional policy is based on hegemonic discourses of what basic problem has to be solved in rural/regional policy and how it is to be solved. The production of these discourses is very much a question of the dynamics of power/knowledge played out between different actors operating at different geographical levels and arenas. In this chapter I will try to show how hegemonic discourses of rural/regional policy are produced and affected by social practice and can be analysed through a study of the relations of power/knowledge. This will be done through two stories. The first is the story of an attempt to construct a regional difference and use this as a policy-instrument, i.e., the construction of a new discourse of spatial division. The second is the story of an attempt to change the hegemonic story of what kind of 'problem' Norwegian rural/regional policy is facing, i.e., the attempt to influence and change the discourse.

The first story raises the question of how the territorial frame of the region can be an alternative and/or supplement to present nation-states, which has been stressed in recent policies in European but also Nordic politics. In the Nordic countries, a region crossing nation-state borders has been used as a kind of policy-instrument in order to promote regional/rural development in scarcely populated areas for several decades. A resent study of such a cross-border co-operation – *The Mid-Nordic region* – shows that the question of success of this kind of policy-instrument is not an easy or simple way to promote regional and rural development (Lysgård, 2001).

The second story is based on an evaluation of a policy-programme set up by the Ministry of regional development and local government, aiming at the most scarcely populated areas in Norway where we also have the strongest tendency of out-migration and population loss. The Ministry has set up a goal for this programme where the main object is not to increase population but to enable the small communities to deal with problems due to population loss. This goal can be interpreted as quite different from what has been the dominating political perspective in Norwegian rural policy since the 1950s – namely to turn the centralizing migration pattern and to preserve/increase population in rural areas (although this still is claimed to be official policy). The programme can therefore be understood as an attempt to create a new discourse of rural policy, containing new ways of understanding the problem in rural areas and new methods of solving the problem. The implementation of the programme has been more or less a success – but not in the sense of the main goal. Why this is so can be analysed as a play of power relations where the result in some sense can be seen as a contest between the 'old' and the 'new' discourse of rural policy (Karlsen et al., 2001; Lysgård et al., 2001).

Power/Knowledge – How to Analyse the Production of Discourses/Narratives

One of the most influential theorists of the power/knowledge debate in contemporary social science is Michel Foucault (1982, 1990, 1994). According to Foucault, knowledge is a strategic or political field, and his project is very much about how the relations between power or strategies on the one side is strongly related to the formation of specific fields of knowledge on the other. Neither knowledge nor power refers to something substantial or essential. Knowledge is a constructed discourse and power is the social relation that makes this discourse recognized as the valid 'truth' at a specific time and place. This means that there exists no such thing or object, which can be called power. Power *is* not – it is exercised. Furthermore, power cannot be seen as something that someone has and someone does not have. Power is a net of social relations, which is unstable and changeable and which expresses a tense relationship between two or more parts. It is never totally on the one or the other side and is in that respect never totally controlled from the one side. Resistance is always a part of the power relations and can be described as counter-power.

To investigate power relations implies an analysis of the relations between the variable forces inside a social and historical specific field. Knowledge becomes the discourse, created out of these power relations, which are dominating at a specific time and place (Foucault, 1982, 1994, 1990; Beronius, 1986; Flyvbjerg 1991). Since power is not an object, we cannot investigate it as an object. Instead we must focus on the specific social practices through which power relations are put into action, because 'power exists only when it is put into action' (Foucault, 1982, p. 219).

The study of power is the study of action and the intentions behind the actions, and therefore we have to put the actors and their actions in the midst of our analysis. In order to structure the analysis of power and production of discourses, Foucault (1982) argues that the actors' actions should be related to the following five conditions. First, *the system of differentiation*, which allows one part to act on other actions. Second, *the types of goals* that are pursued by the actors. Third, *the means* that are used to establish a power relation and put it into function. Fourth, *the forms of institutionalization* that are created or changed by the power relations. And fifth, *the degree of rationalization* and effectiveness concerning instruments used, the obtained results and the costs of the actions.

The intentions of the involved actors and concrete actions are what make the dynamics of power/knowledge in the production and reproduction of the discourses of regional/rural policy, and they can be summarized in two conceptual axes. One is an expression of the degree and level of the concrete action in relation to the co-operation and goes from total *indifference* to active *intervention*. The other is an expression of the dynamics in the different actors' intentions and normative positions towards the co-operation and goes from *counteraction* to *support*. When these two conceptual axes are crossed, they form a *field of power dynamics* (Lysgård, 2001). By analysing the actors' intentional actions in relation to others' actions in the production of the discourse(s), it becomes possible to understand how and why the discourse is made and how and why the intended discourse(s) of regional/rural policy became a success or a failure.

Figure 6.1 The field of power dynamics

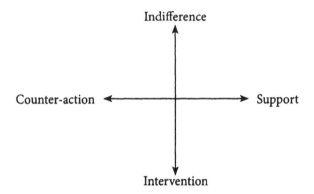

Regionalization as a Policy Instrument – The Mid-Nordic Region

The Region in Rural Policy

The development of a regional level is not only a result of recent EU politics. Functional, cultural and political regional units, crossing nation-state borders, can be traced several decades back in history. In the Nordic countries several cross-border co-operations have existed since the early 1970s as a part of the common Nordic policy in the Nordic Council of Ministers.[1] The motive for these regions vary from making a stronger economic and functional territorial integration, to a revival of an 'imagined' common cultural/ethnic base; but the basic motivation is to create a territorial division that seems to be better suited to achieve political goals than the existing administrative borders and functional structures.

Cross-border co-operation can therefore be understood as an instrument in the construction and implementation of regional policies. Several of these cross-border regions are relatively scarcely populated, even in a Nordic sense, and have been created from the idea that through joint political pressure they can compete with the strong attraction of the most urbanized areas (especially the capital cities) on economic/financial activity and skilled labour. The cross-border regions can therefore be seen as instruments, not only for making and implementing regional policies, but also for *rural policies*.

How then can a region become an instrument for doing regional/rural policy? The region as a spatial category has for ages been an important part of the political discourse as well as of the academic discourse, and also as an identifying aspect in everyday life of ordinary people. But the use of the concept has been characterized by the widely held misconception that the region is some kind of natural essential entity. In a political sense this has led to the belief that there is a natural relation between race, ethnicity, nationality and territory (Peet, 1998). The idea of the natural, cultural homogeneous region with a real existence in an essential sense is therefore central in the study of the political meaning of the concept of region. This idea must not be mixed with an ontological understanding of the concept of region, but as a strategic position in the production of policies and in the everyday production and reproduction of regions and regional identity (Lysgård, 2001).

Space and Power/Knowledge – Space as Discourse and Region as Narrative

In order to understand the region in this case, it is necessary to make a short excursion into the concept of space. Space is a social dimension under constant construction and transformation. Space is produced through social practice and is at the same time both a means of production and a result of social action (Soja, 1989; Lefebvre 1991; Simonsen, 1996; Werlen 1993). This notion of space has not been the ordinary position within either human geography or in any other academic field dealing with space. In fact the most common position is that space is the geometry of human life. Space is regarded as the form or the container-like entity where social action takes place. The problem with this position is that it takes all meaning out of the spatial dimension and reduces it to plain geometry or localization in absolute space. My intention is instead to regard space or spatiality as a social dimension: '(Spatiality) captures the ways in which the social and spatial are inextricably realized one in the other; to conjure up the circumstances in which society and space are simultaneously realized by thinking, feeling, doing individuals and also to conjure up the many different conditions in which such realizations are experienced by thinking, feeling, doing subjects' (Keith and Pile, 1993, p. 6).

One of the most significant theorists promoting the idea of the spatial dimension as a social product has been Henri Lefebvre (1991). He claimed that space and place actually have an ontological status since both are embodied in human activity. Thus space is not something substantive or objective in a material or physical way. In order to analyse the spatiality of social activity, Lefebvre operates with different 'fields' of space in a spatial triad: *spatial practice, representations of space and spaces of representation*; a triad referred to and used in a multitude of contemporary geographical studies (Soja, 1989, 1996, 2000; Harvey, 1990; Simonsen, 1991, 1993, 1996; Merrifield, 1993; Gregory, 1994; Pile, 1996; Pløger, 1997; Peet, 1998; Shields, 1999; Lysgård, 2001).

Social practices, or *perceived spatiality*, structure daily life through people's perception of social interaction. The perceived space is produced in the relations and dialectics between institutional systems and daily experiences and practices and is mainly oriented around the concrete materiality of spatial forms, or things that can be empirically mapped. *Representation of space*, or *conceived spatiality*, is the conceptualized and constructed space produced by professionals and technocrats, which allows us to talk about and understand the material space both within everyday life situations and different academic disciplines, mediated through systems of verbal, intellectually worked-out signs. *Spaces of representations*, or *lived spatiality*, are produced by the everyday life of inhabitants and users. This is space as directly lived and experienced through its associated symbols and images used by people in everyday practice and is a way of escaping from the illusion of an ontological and epistemological separation between objectivism and subjectivism.

Social spatiality comes out of a process of production, including social practice, which with its background and influence from historically produced spatial representations (the conceived space) constructs and rationalizes the established spatial representations and discourses (the lived space). This influences the practice of production through specification and adjusting the movement of the body, the materiality of society and the spatial relations between subjects (the perceived space). The division between the three spatial aspects must therefore mainly be seen as an analytical division that has the potential to form an epistemological approach towards the study of social spatiality.

Regions are results of such processes of social practice involving production of spatiality in all the three Lefebvrian forms. The region as a spatial category is, in other words, a social product that can have different meanings in different situations, and one and the same region can in different situations and for different actors take different roles and represent different intentions/goals. It is therefore meaningless to define or make regional typologies as some kind of absolute essential existences, as often done by geographers and other social scientists. It is more fruitful to try to identify the aim/intention lying behind a regional division and construction and how this division into regions gives meaning to and implication for the spatial organization of society.

The region can be an important tool for several scientific disciplines and others in order to systematize, classify and analyse data: *the methodological region*. The region can also be an important aspect in the way people relate themselves to the outside world and the process in which people identify with an imagined community: *the identifying region*. And thirdly, but maybe most important here, the region can also be an important tool in policy making and in defining the geographical target areas of policy. Using the region as a policy instrument enables politicians and planners to define which areas need change, demand planning, and which areas should be affected by regional/rural policy.

The Story of the Mid-Nordic Region

In order to cope with a situation of scarcely populated areas and severe population loss, to maintain or increase employment possibilities, and secure settlement and living conditions in these areas, one strategy for several counties in the Nordic countries has been to establish regional co-operations with counties in a more or less similar situation across the nation-state borders. One region that is more or less based on such argumentation is the *Mid-Nordic region* in the middle of Scandinavia. It was officially launched in 1978, after an initiative taken by a few chambers of commerce three years earlier. The region is located mainly between the 62nd and 64th latitude, and is put together by regionally based political administrations in Norway, Sweden and Finland. The first years of the co-operation were mainly oriented towards the strengthening of opportunities for doing business (trading, tourism, energy and industry) and improving communication (transport) in the area. The capital areas (Oslo, Stockholm and Helsinki) had (and are still having) a strong attraction on both finance and skilled labour and there was a fear that the centralizing forces towards the capital areas would affect employment possibilities in a negative way and cause further population loss.

During the first years of the co-operation the main activities were oriented towards creating an economic basis for the maintenance of employment and settlement. The cultural part became more and more important during the late 1980s and 1990s. In the strategic activity plans of the 1990s, for instance, culture and the development of a regional identity common to the inhabitant of the region became an issue. From a focus on business and communication in the early 1980s, the vision and main goal formulated in the latest strategic plan also focus on securing a sustainable development and protecting the environment and developing a sense of historical and cultural belonging among inhabitants (Lysgård, 2001).

The aim of the case study was to find out if the short history of the Mid-Nordic co-operation had produced some kind of substantial foundation, which allows it to develop as a legitimate and recognized region and to become a part of everyday practice for both people who live in and/or outside its defined boundaries. The analysis shows that the narrative of the Mid-Nordic region has developed as a legitimate discourse only for very few people but not as a part of everyday life in the areas. The answer to the research question is therefore a *conditional* no. There seems not to have been developed any significant materialized/institutionalized spatial practices of by and for co-operation and interaction in the area. The presentation and marketing of the 'image' of the region as a political co-operation and/or unitary region in a physical, social and cultural sense, is not particularly effective and is received by very few people. A regional consciousness about the Mid-Nordic region hardly exists, and the co-operation seems not to play any part on the regional arena neither in the Nordic countries nor in a European context. The answer is a *conditional* no, only because it is analysed from of a time-space specific context, which is in constant change, and even if it does not seem likely, a future situation with production of such a substance cannot be precluded.

Power/Knowledge in the Production of Spatiality

The Mid-Nordic region is limited by a weak and missing production of spatiality. Neither perceived (spatial practice) nor lived space (spaces of representations) are produced or reproduced to any amount. The Mid-Nordic region exists mainly as conceived space (representations of space), but since this space is only known by relatively few, spatial production in this sense is not capable of developing the Mid-Nordic narrative as a legitimate and recognized regional entity. Reproduction and further institutionalization of the Mid-Nordic region as perceived and conceived space will not be manifested as a concrete spatiality without becoming a part of everyday practice. The everyday practice of people in the area works more or less in opposition to the conceived space, presented in future plans for the co-operation. In everyday practice other spatialities dominate and are used in identification and spatial organization of social life. Lived spatiality is thus the most important expression of power that counteracts the Mid-Nordic co-operation. However, this is not an active or conscious opposition, but appears as *indifference* to the perceived and conceived space. Both political-administrative actors, actors that through their activity could establish, develop and use such spatial structures, and the general population in the area, display such indifference.

According to Foucault's five conditions we *first* find that the Mid-Nordic region as a system of differentiation, understood as an attempt to construct a spatial difference, is not very successful. The problem is that the regional narrative to a very limited degree is perceived or lived, since very few actors recognize the narrative of the region as a legitimate geographical entity different from other known regional demarcations. *Second*, the goal of the co-operation is not in harmony with the dominating practices and spatial representations in the areas, and therefore, relatively few actors support the objective of the co-operation. *Third*, the Mid-Nordic co-operation has very few available means in order to obtain the objectives, when it comes to political authority, economy and credibility among the inhabitants. *Fourth*, the region is characterized by lacking institutionalization and the co-operation is therefore supported to a limited degree only by established formal and/or non-formal institutions and institutionalized functional or cultural practises. And *fifth*, the degree of rationalization is relatively low, since the means used do not seem to be very effective in dealing with or relate to the system of differentiation or in developing an institutionalized substance for the discursive existence of the region.

Domination in the field of power dynamics can be localized in the categories of *counteraction-intervention* and *counteraction-indifference*, where the first one is the most dominant. The biggest problem in the development of the Mid-Nordic region is mainly found in the dominating position that actors in this category have in relation to actors who seeks to intervene and support the co-operation as an institutionalized region. The Mid-Nordic region cannot be established and presented as a legitimate and recognized narrative, because the agents involved lack sufficient strength and authority in relations of power/knowledge.

In the field of power dynamics we can identify the main problems and constrains for using the regions as policy-instruments for regional and rural development. The study of the Mid-Nordic region reveals at least three significant disharmonies, which often become apparent in this type of policy making. First, the political region often does not coincide with political intentions in other arenas and political levels. Second, the political region often does not coincide with the development of a globalized economy based on networks with few geographical limitations. Third, the political region often does not coincide with the everyday production of spatiality and identification of ordinary people living in or outside the area.

Changing the 'Problem' Discourse of Norwegian Rural Policy

Rural/Regional Policy in Post-War Norway

This second story will focus on an attempt to change the hegemonic discourse of the basic problem of Norwegian regional/rural policy. In Norway the research field 'regional development' is regarded almost as synonymous with research on 'rural development.' Regional political discourses have likewise a rural profile, and have been focused on rural problems during almost the entire post-war period (Berg and Lysgård, 2001).

Norway and the other Nordic countries, except Denmark, have vast areas with extreme low population density. These areas have been the focus of Nordic regional policy during most of the post-war period up to relatively recently. Nordic regional policy in the post-war period has therefore been what some researchers have called 'periphery policy' (Mønnesland, 1997), or perhaps better 'rural policy.' Although in Norway 'urban' and 'rural' are not used in the same way as in central and southern Europe, it is often more relevant to talk about 'rural policy' than 'regional policy,' because of its rural focus.

The production of the hegemonic discourse of rural policies, in which the idea of stabilizing and maintaining the settlement patterns has been in the forefront, can be traced back to the early 1950s. After the second world war, Norway faced the task of rebuilding the country. It was almost undisputed that the benefits should be regionally distributed, and that a regional policy was needed for this purpose. *'Town and countryside – hand in hand*' was a widely accepted slogan in the 1950s (Teigen, 1999). Regional balance was set up as a goal, and poorly situated regions had to benefit.

In the 1960s. a nation-wide bureaucracy to administer regional policy was establishment. Rationalization processes in agriculture caused unemployment in most rural areas, which in turn caused migration to urban areas and an urban growth. Norway thus faced a situation with severe problems in rural areas, and regional policy was increasingly meant to stimulate economic growth outside defined congested areas. With the idea of even economic growth in all parts of the country, another idea was connected in the early 1970s, namely that of stabilizing the settlement pattern. This idea has permeated Norwegian regional policy ever since. The goal was well on its way to be reached during the 1970s, since there was a decrease in net out-migration from peripheral regions and a decrease in in-migration to central regions; in other words a relative stabilization of the pattern of settlement.

In the early 1980s a new wave of migration to central regions, mainly caused by employment opportunities, appeared. The objective of stabilizing the settlement pattern became as important and as difficult as ever before. In the late 1980s there were, however, tendencies towards stagnation and employment problems in most parts of the country and the overheating problem in urban areas – if not the capital area – was reduced remarkably. Whole regions with a cluster of involved firms experienced structural problems as a result of industrial recession (Mønnesland, 1997). Again a reassessment of regional policy, not least its geographical profile, was needed. In a situation with new nation-wide problems rather than specific rural or urban problems, more responsibility was delegated to lower regional levels and the attention was turned to local entrepreneurship rather than redistribution. Among other things, a programme for the development of private service firms was established to contribute both to employment and service supply.

While the first half of the 1990s can be characterized as a period of relative stability as regards the settlement pattern, the second half of the decade saw an alarming increase in migration from peripheral regions. A big difference compared with the beginning of the decade was that larger centres in the more remote regions, which used to balance the total figure of the region, were increasingly faced with migration loss (Hanell, 1998).

A Rural Policy Programme – 'Utkantprogrammet'

Towards the end of the 1990s there was, therefore, again a shift in the focus of regional policy, described by the Minister of regional development and local government (March 1998) as 'a shift from a policy directed towards all regions towards a policy with a stronger focus on the weakest parts of the country' (Aalbu, 1998, p. 9). In the latest Review to Parliament in 2001 this policy is continued but now alongside the idea of the so-called 'robust' region. In order to change the focus on maintaining and stabilizing settlement patterns and stimulate economic growth in the weakest parts of the country, a programme (*Utkantprogrammet*) was set up in 1997 with the goal to help the weakest parts of the country to better cope with out-migration, centralization and population loss and a new focus on total welfare instead of the one-sided goal of economic growth.

This means that the discourse produced through fifty years of political history was now challenged by the ideas of this programme. And the question is whether this programme can make such changes in the hegemonic discourse of what is the Norwegian rural 'problem.' The success or failure of the implementation of this programme can therefore be analysed in more or less the same way as the success or failure of the region as a policy instrument. The programme runs out in 2001, but is prolonged a part of official rural policy under the slogan *small communities* (*småsamfunn*) along with the slogan *robust societies* (*robuste samfunn*).

Power/Knowledge in the Changing Discourse of Norwegian Rural Policy

The programme has in itself been relatively successful because the activities in the programme have been carried through more or less as planned. A more interesting question is if the programme actually has achieved its main objectives: that of moving the discourse of the 'problem' away from a focus on preserving and increasing the population in the rural areas. From this point of view the programme has more or less failed.

According to Foucault's five conditions we *first* find that the programme has created a system of differentiation between those communes who are to be defined as *utkantkommune* and those who are not. On the one side the concept *utkant* (i.e., rural periphery) can be seen as something negative – it is stigmatizing: the communes of 'hopelessness.' On the other side to be defined as a *utkan* commune is seen as something positive, because these communes receive help from the government. But this aspect also creates the impression of unfair treatment among local governments not participating. As a result of the negative connotation of the concept *utkant*, a new concept is introduced – *småsamfunn* (small communities).

Second, some of the involved actors follow the original objective of the programme and try to develop a new focus in the rural and regional policy, i.e., a shift from economic growth and stabilizing/increasing population to increased welfare for those staying behind. Some do not care about the objectives and are just using it as a way of financing local activities planned anyway. Others again have reformulated the objectives to fit into the local political context where increasing population still is the only legitimate political objective. This last group is significant and has an important influence on the possibility of changing the discourse, because most communes involved find the meaning of the programme objectives intolerable and impossible to mediate in a local political context.

Third, financial support to projects and specific actions is not new in Norwegian rural policy and for some of the communes the programme has only been a financial supplement to already planned activities. The really new aspect can be found in the focus on the promotion of the image of rural communes as good places to live. Several of the communes have started to make internal and external place marketing and image improvement activities in order to increase pride and satisfaction of the home place and attract lifestyle migrators.

Fourth, the programme has more or less cut off parts of the traditional administrative institutions by letting the communes communicate directly with the ministry, without involving the county administration as would be the ordinary practice in regional policy. This has been a popular move from the rural communes participating in the programme, but not so popular seen from the county administration side. Now that the ideas from this programme are becoming the focus of regional/rural policy, it is actually planned that the counties will administer it.

Fifth, the degree of rationalization is somehow reduced by the fact that the concept used has actually stigmatized the rural communes as the communes of hopelessness when the intention actually was quite the opposite. It is moreover reduced by the different understandings and practice of the programme objective. And last, it is reduced by the obscurity of the way in which the programme has been related to existing administrative institutions.

As an attempt to change the focus of Norwegian regional/rural policy, the programme has not been very successful. Although the objective has been communicated and agreed on by politicians at the national level, the main counter-power working against this process have been politicians at the local level. They have not been able to take the new focus as their own policy because they would commit political suicide. The resistance to changing the focus has not been directed against the idea that the weakest parts of the country should be better capable of dealing with out-migration, centralization and population loss and of a new focus on total welfare, but against the one-sided goal of economic growth. The problem has been that for a lot of people this also means that they had to let go the objective of maintaining the settlement pattern and increasing population in rural areas. And this is impossible for local politicians to communicate to their voters.

No one really works against the programme as a direct *counteraction*, but several actors, especially local actors within the involved rural communes, have adapted the programme objective to their own political agenda and are more or less *indifferent* to the programme objectives themselves. This has been the major expression of power working against changes in the discourse of the Norwegian rural policy. This situation introduces a third dimension to the field of power dynamics. In the dimension of intentions for action, practical politics separates into at least two layers dedicated to different political arenas. One intention is expressed at the national arena, while another is expressed at a local arena, and maybe we also find a third that is not expressed at all, but anyhow guides the action.

Conclusion

The two cases show that both the process of defining what are the rural problems and the actions suitable to solve the problems, are time and space specific processes of social construction. Rural policies are by all means discursive constructions that can be analysed within the relation between power and knowledge. The success or failure of rural/regional policy actions can be understood as expressions of power and counter power. The dynamics of the power relations involved in the construction of rural policy can be analysed in the tension between two conceptual axes. One is an expression of the degree and level of the concrete action in relation to the co-operation and goes from the total *indifference* to active *intervention*. The other is an expression of the dynamics in the different actors' intentions and normative positions towards the co-operation and goes from *counteraction* to *support*.

Both cases are about the process of changing the discourse of rural policy, and the conclusion of both cases points at the fact that changes in discourses cannot be made only by decisions by a limited number of politicians at a certain geographical level. The changes do not appear as real changes before they are recognized and legitimized by a broad spectre of actors involved in the process as a valid aspect of the hegemonic discourse. A significant source of power in both cases is actually not powerful politicians at the national level, but actors at the local level, the rural population. The slightly underestimated expressions of power that becomes visible through this analysis is on the one side the direct counteraction to changes in rural/regional policy at the local level, but another important expression of power with a deep impact on the possibility of changing the discourse is the indifference shown by those people whom the rural/regional policy is actually about. The point to be made is that to make real changes in the discourse of the rural, one needs involvement from the people it is about.

Note

1   The Nordic Council of Ministers (established in 1972) is a co-operation for the Nordic countries governments and its task is to implement the policies of the Nordic Council. The Nordic Council (established in 1952) is the co-operation between the parliaments of Norway, Sweden, Finland, Denmark and Iceland.

References

Aalbu, H. (1998), The pendulum of regional policies in Norway and Sweden. *North* 9 (1) pp. 7–10.
Berg, N.B. and Lysgård, H.K. (2001), Rural development and policies – the case of post-war Norway. In K. Halfacree, I. Kovach and R. Woodward (eds), *Leadership and Local Power in European Rural Development*. London, Ashgate.

Beronius, M. (1986), *Den diciplinära maktens organisering. Om makt och arbetsorganisation.* Lund, Arkiv avhandlingsserie nr. 23.

Flyvbjerg, B. (1991), *Rationalitet og magt. Bind I Det konkretes videnskab.* Danmark: Akademisk Forlag.

Foucault, M. (1982), The subject and power. In H.L. Dreyfus and P. Rabinow eds, *Michel Foucault: Beyond Structuralism and Hermeneutics.* Brighton, The Harvester Press, pp. 208–226.

Foucault, M. (1990), *The History of Sexuality.* Vol. 1 *An introduction.* London, Penguin books.

Foucault, M. (1994), *Overvåkning og straff. Det moderne fengselssystems framvekst.* 2. Edition. Oslo, Gyldendal Norsk Forlag.

Gregory, D. (1994), *Geographical Imaginations.* Cambridge/Oxford, Blackwell.

Hanell, T. (1998), Nordic peripheries in trouble again. *North* 9 (1) pp. 11–15.

Harvey, D. (1990), *The Condition of Postmodernity.* Cambridge/Oxford: Blackwell.

Karlsen, J., Langhelle, O., Ryntveit, A.K. and Lysgård, H.K. (2001). *Følgeevalueringen av Utkant-programmet, årsrapport for 2000.* Prosjektrapport 51/2001. Kristiansand, Agderforskning.

Keith, M. and Pile, S. (1993), Introduction part 2: The place of politics. In M. Keith and S. Pile (eds), *Place and the politics of identity.* London/New York, Routledge, pp. 22–40.

Lefebvre, H. (1991), *The Production of Space.* Oxford/Cambridge, Blackwell.

Lysgård, H.K., Karlsen, J. and Ryntveit, A.K. (2001), *Endring og læring i et distriktspolitisk eksperiment – En teoretisk tilnærming til evaluering av Utkantprogrammet.* Fou-rapport nr. 11/200. Kristiansand Agderforskning.

Lysgård, H.K. (2001), *Produksjon av rom og identitet i transnasjonale regioner – Et eksempel fra det politiske samarbeidet i Midt-Norden.* PhD. thesis, Trondheim, Department of Geography, NTNU.

Merrifield, A. (1993), Place and space: a Lefebvrian reconciliation. *Transactions of the IBG* 18 (4) pp. 516–531.

Mønnesland, J. (1997), Regional policy in the Nordic Countries. Background and Tendencies 1997. NordREFO, 2.

Peet, R. (1998), *Modern Geographical Thought.* Oxford, Blackwell.

Pile, S. (1996), *The Body and the City. Psychoanalysis, Space and Subjectivity.* London/New York, Routledge.

Pløger, J. (1997), *Byliv og modernitet – mellom nærmiljø og urbanitet.* NIBRs Pluss-serie 1:97. Oslo, NIBR.

Shields, R. (1999), *Lefebvre, Love and Struggle. Spatial Dialectics.* London/New York, Routledge.

Simonsen, K. (1991), Towards an understanding of the contextuality of mode of life. *Environment and Planning D: Society and Space* 9, pp. 417–431.

Simonsen, K. (1993), *Byteori og hverdagspraksis.* København, Akademisk Forlag.

Simonsen, K. (1996), What kind of space in what kind of social theory? *Progress in Human Geography* 20 (4) pp. 494–512.

Soja, E.W. (1989), *Postmodern Geographies. The Reassertion of Space in Critical Social Theory.* London/New York, Verso.

Soja, E.W. (1996), *Thirdspace: Journeys to Los Angeles and Other Real-and-Imagined Places.* Oxford, Blackwell.

Soja, E.W. (2000), *Postmetropolis. Critical Studies of Cities and Regions.* Oxford, Blackwell.

Teigen, H. (1999), *Regional økonomi og politikk.* Oslo, Universitetsforlaget.

Werlen, B. (1993a), *Society, Action and Space. An Alternative Human Geography.* London/New York, Routledge.

Chapter 7

Constituent Power of Dutch and EU Discourses on Rurality and Rural Policies

Henri Goverde

A power perspective can be useful in explaining why the administrative capacity and performance of specific new policy arrangements and political institutions concerning rural innovation and development have been a success or a failure, both domestically and internationally. From this perspective, three layers of power are relevant: 'power over', 'power to' and 'shifts in power relationships'. This chapter investigates the constituent power of Dutch discourses concerning rurality and EU discourses concerning rural policies. The central hypothesis is that the more congruency there is between these discourses at the domestic and international levels of governance, the better the policy arrangements will be in both rural politics and rural development.

Discourse analysis will be the main research approach used here. This approach will be used in order to discover the relevant changes that have taken place within the different layers of power in the field of rural policy and development. The discovery and the descriptive analysis of a discourse help to explain why a certain perception of a problem (for example consumer concerns about livestock diseases) tends to become dominant and authoritative, whereas other points of view receive little or no support. Discourses and discursive action are sources of power and because of that a 'discourse' has 'constituent power'. Such power is formative or rather essential because it gives meaning to our world (social environment). From this perspective, all types of policy categories are not only instruments for realizing public goals, but they can also be perceived as symbols which give meaning to, for example, 'rurality' in specific areas.

In a set of case studies a few Dutch authors (Frouws, 1998; Hidding et al., 1998; Boonstra, 2002) have reconstructed relevant discourses concerning 'thinking and talking' by academics and policy makers *about* rurality and 'thinking and talking' by people *in* rurality in the Netherlands. This chapter first aims to review these scientifically constructed discourses. Secondly, the (main shifts in) EU policies since the Cork Declaration (1996) concerning rural development will be reconstructed. Thirdly, it will be investigated whether there is a congruency or a contradiction between the dominant political EU discourses and the (dominant) discourses concerning rurality in the

Netherlands. In a specific section, the data will be interpreted from the perspective of different types of power: resource-based episodical power ('power over'), dispositional power ('power to') and structural/systemic power ('shifts in power balance'). Attention will be paid to the question of whether or not these discourses are gendered. Of course, it is reasonable to expect the long-term continuation of the policy network of the EU- and member state rural institutions. Subsequently the question arises of 'how do congruent or competing discourses influence local governance in rural areas, particularly during catastrophes such as the recent epidemic livestock diseases?'

The empirical research has been organized around the following *central hypothesis*: The more congruent Dutch and EU discourses concerning rurality and rural development are, the more constituent power these discourses will have. The research method used for this chapter has been confined to desk research. The results are based on scientific literature, articles in newspapers and (weekly) magazines, policy documents and information about different (governmental) institutions, for the most part collected from the Internet. The data collection was completed by the end of 2001.

Discourses about Rurality in the Netherlands

In this section the main topics are presented in two series of current discourses about rurality in the Netherlands. The construction of these discourses emphasizes a relevant methodological issue about the theoretical relationship between 'discourse analysis' and 'the structuration of policy issues'. This issue demonstrates the political relevance of congruency or incongruency between discourses supported in policy-making at divergent governmental levels and discourses enhanced in 'everyday life.'

Three Coherent Sets of Thoughts about Rurality

In a recent essay on 'the politics of rural development in the Netherlands' Goverde and de Haan (2002) refer to three main categories of rural discourse, which Frouws (1998) distinguished as being: the agri-ruralist, the utilitarian and the hedonist discourse. These themes can be summarized as follows (Goverde and de Haan, 2002). The *agri-ruralist discourse* emphasizes the central role of farmers in the countryside renewal process. The values of the countryside, such as the landscape, open space, natural resources and heritage, have always been based on the co-production between man and nature, and farmers have subsequently played an essential role in preserving this identity and in meeting the social demands for maintaining an attractive countryside and high-quality food products. Self-regulation and endogenous development are of essential importance in the revival of rural areas and the establishment of a new identity between local actors and their environment. This discourse concerning rurality is a modern version of the glorification of autonomous rural communities, which have now opened up to serve the needs of urban society.

In emphasizing the *social dimension*, this way of thinking differs from the *utilitarian discourse*, which focuses entirely on the *economic dimension*. The protection of the countryside obstructs opportunities for taking advantage of its economic potential. By completely integrating rural space in the dynamics of modern markets for housing, recreation, quality food, high-tech agriculture, attractive landscapes and so on, it can better meet society's growing demands for space to live, work and recreate. This neo-liberal discourse believes that the commoditization of rural areas and the breaking down of artificial boundaries between city and country are the best guarantees for meeting the diversity of social needs for quality. According to Frouws, this discourse represents a long-standing Dutch tradition of reclamation, exploitation and commercialization of rural land.

The third discourse mentioned by Frouws is the *hedonist discourse*. Here the emphasis is on the *cultural dimension* of the countryside, in particular its scenic beauty and natural values. As such, it plays an important role as a counter-image to congested urban life. This discourse is rooted in nature conservation movements and elite notions of the countryside as the garden of the city. The role of the state and derived forms of local governance are mentioned as the most important regulative mechanisms.

Methodological Intermezzo

The discourses presented here are not based, as Frouws has also admitted, on the representation of a real division in interest groups or parties. These discourses should therefore not be interpreted as an array of programmatic choices in the field of rural politics. They should be seen as an analytic construction of rather coherent sets of thought (*about* rurality). The presentation of these discourses can be read as a collection of themes that may co-exist in different combinations (*in* rurality). Discourses are both results of interaction as well as resources and rules in interaction. Of course, it can be expected that the synchronic existence of different discourses at the macro level can produce problems in the praxis of policy-making at all levels of governance. These discourses can inspire projects that compete in spatial, social and environmental impacts, that are related to different interests, and as such result in the continuation or the contestation of power relations.

Recently Boonstra (2002) formulated some criticism of the discourse categories constructed by Frouws and others.[1] Boonstra agrees with the empirical observations made by Goverde and de Haan (1999), who argue that all discourses believe in the human capacity to control nature as well as social development, which they consider to be a typical Dutch characteristic. However, she argues that all discourses are constructions by researchers which cannot be read directly from rural praxis. Therefore, she prefers a research method that can relate discourses with social praxis and power.

Indeed, from a methodological perspective rural development should not be read as a 'text' only (as in the narrative tradition) but also 'in context'. The discursive variety can be analysed in processes of decision-making about rural development at the regional level. On the one hand, these (policy) processes are influenced by societal and scientific debates. On the other hand, regional processes of policy-making are still close to the way of thinking and the practices found in everyday life in the rural areas. This results in the assumption that challenges (opportunities as well as threats) produced by discursive variety can be discovered in such a way that relevant information for policy development can become available.[1]

Four Policy-Relevant Discourses

Boonstra (2002, pp. 248–49, 253) proposes the following typology of policy-relevant discourses in Dutch rural areas, based on empirical discourse analysis elaborated on recent projects concerning rural innovation in North-West Frisia (the only LEADER-area in the Netherlands) and 'de Graafschap' – an area labelled as a landscape of special cultural value and natural beauty.

(Neo-)modernization discourse. Definition of the problem: during the period 1945–1980 the key issues were social-economic backwardness of rural areas, relatively low productivity, and low contribution to the national economy and welfare. After 1980 new issues entered the arena, such as (ecological) sustainability, food-quality risks, and animal welfare. Suggestions made for solving problems: increased productivity, production and export, integration of rural activities in an urban society, division between urban and rural spaces. Actors: farmers distinguished as those 'who leave' and those 'who stay' (agrarian entrepreneurs); farmers' elite functions as mediator between government and farming population.

Wilderness discourse/wasteland discourse. Definition of the problem: decline of biodiversity, fragmentation of ecosystems. Suggested solutions: minimal human interventions, self-regulating nature, development of related system of existing nature preservation areas and newly developed nature areas (ecological infrastructure). Actors: primarily professional elites in fields like ecology, nature protection and physical planning.

Multi-functional space discourse. Definition of the problem: decline of rural cultural values; rural areas in general and agriculture in particular do not fulfill the needs of society; strategy of spatial zoning comes to an end because of the lack of space. Suggested solutions: mutual development of cities and the countryside, multifunctional use of available space, integration of functions, diversification of rural economy, environmental management. Actors: farmers, interest groups, rural entrepreneurs as well as the population in general. These groups pursue not only their private interests, but they also have responsibility for more common interests. The government acts in the general interest, looks for failures of the market, and creates conditions for self-government and well-functioning markets.

Discourse concerning the establishment of new business sites (vestigingsplaats). Definition of the problem: underexposure of the economic potentialities of the countryside and shortness of market-integration. Solutions proposed: optimizing added value per acre, enlargement of free choice for citizens and firms, market integration and market functioning, fewer obstacles for land use, better accessibility. Actors: citizens (consumers) as well as firms try to maximize utility. Public actors are completely dependent on initiatives of private actors.

Boonstra does not distinguish *a hedonist discourse* in Dutch rural practice. Besides scenic beauty and the presence of the nature conservation movement, the emphasis on natural values in farming seems rather underexposed. In fact, a recent report of the National Advisory Council for the Rural Areas confirms that 'organic farming' in the Netherlands is not very well developed in comparison with several other Western European countries.[3]

Dutch Discourse Concerning Epidemic Livestock Diseases in EU Context

The discourses mentioned above do not always explicitly take into account the relationship between the national policy envelope and the EU policy system. A discussion in the Dutch Parliament (21.3.01/TK no 21501–16/290) related to the ongoing crises in animal health (Swine fever, BSE, Foot and Mouth Disease) revealed the crucial elements in the Dutch discourse concerning EU politics related to agriculture and rural development. Generally speaking, the political actors are keen enough to relate measures for crisis management to the restructuring of the sector in the long term. Such a debate also illustrates how international politics are constantly influenced by domestic variables. The following statements are characteristic of the debate:

- Farmers receive consistently lower prices at the slaughterhouses, whereas the consumer prices in the supermarkets stay at the same level or they increase slightly. The EP amendment to change the buying-up rules from a 70 percent EU payment to a fifty-fifty regime implies a shift in financial regime, i.e., from *finance communautaire* to a system of 'co-financing'. This can produce a re-nationalization of rural policy in the long term. It seems that the EU uses the livestock disease crises both to decentralize many costs to the individual farmer as well as to promote the liberalization of the market. Both tendencies imply that the whole crisis demands permanently higher budgets or they will turn out to be a dead-end street for the sector.
- European society ought to be offered a crisis-management approach based on solidarity. The revision of the tax system should ensure that consumers, just like other actors in the agri-business chain, also share responsibility for the solution of the crises as well as for the restructuring of the sector.
- From an ethical point of view, it is amazing and unacceptable that hundreds of thousands of cattle will be destroyed, even though they are in good health. It is also unacceptable that EU premiums are related to a buying-up arrangement. This method is absurd, because it accepts that farmers breed only for buying-up.
- The so-called 'seven-point plan' of the EU Commission (see below) accepts a diminishing demand of 10 percent for the long term. Therefore, an extra evaluation

of the EU Common Agricultural Policy (CAP) is unavoidable. However, any change in the CAP should be based on an area-oriented approach, independent from individual producers.

- The EU permission to start organic farming on fallow land is only 'window dressing.'
- During this type of crisis further breeding of cattle should be forbidden.
- In order to reconquer consumers' trust, meat from non-EU countries which has not been inspected should not be imported while European meat is being controlled under strict criteria. However, an internal European market prohibits a Dutch *Alleingang* in closing its borders to meat from non-EU countries as well.
- The EU proposals for agriculture and rural development should be evaluated with respect to the following seven fields: food safety, consumers' concern/trust, the French-German political relationship, the worldwide perspective for agriculture, the price of a litre of milk directly from the farm, radical stop of premiums for bulls, and competition regime (to attack turbo-capitalism).
- It is not in the interest of the Netherlands that French and German agricultural approaches differ widely: in Germany there is currently a movement to change the policy towards supporting a more organic, extensive way of farming, whereas in France the politicians, who are currently in an election period, are promising to help the farmers with extra money from the national budget.

Conclusion

Recent literature concerning discourses about rurality in the Netherlands exposes a rather impressive 'discursive variety.' This textual discursive variety should be made meaningful as challenging policy information in context. As soon as the different discourses are linked to and become congruent with the legitimately recognized policy problems in the political arena, the (ex-post) conclusion can be that these discourses received constituent power. Particularly, the Dutch discourses mentioned above all received some constituent power. The 'agri-ruralist discourse,' the 'multi-functional space discourse,' the 'discourse concerning the set up of new economies,' and the 'wasteland discourse' (the latter combined with the 'hedonist discourse') all mobilize some power resources to rearrange in one way or another the dominant paradigm of modernity. However, the '(neo-)modernization discourse' and the 'utilitarian discourse' still seem to reflect the dominant frames in the rural development project.

Reconstruction of EU Rural Policy Discourse since the Cork Declaration of 1996

It can be argued that the Cork Declaration (1996) was a political statement that recognized the urgency to adapt the CAP to, on the one hand, a new transnational political-economic context and, on the other hand, to new domestic priorities in various member states. The Cork Declaration reflects the idea of accommodating the modernization project in rural Europe to new concerns of interest groups, governments and individual consumers. In fact, it was the beginning of an end to a pure firm-oriented economic approach in rural development.

CAP *in Historical Perspective*

Before the EEC was constituted each country had its own agricultural policy. No coherence could be observed between the different agricultural policies of the different nation states. After the Treaty of Rome (1957) the EEC member states organized a meeting, led by the Dutch EC representative Sicco Mansholt, to constitute a common agricultural policy (CAP). A long preparation period was needed (1957–1964) before an interim period was started. In 1968, the European Council agreed that the time had come for a policy with one target price and one intervention price. This subsection will focus on the core of the CAP as well as the reasons and the urgency to reform it.

The CAP '... lies at the heart of the practical application of the European vision ... Enshrined in Article 39 of the Treaty of Rome in 1957, the commitment to common organization of the market for food survived intact and was incorporated unamended into the Maastricht Treaty in 1992. The other common policies do not exercise such symbolic power within the Union. (...) The totemic role of the CAP is enhanced at a time when the new elements of European unity outlined at Maastricht, the common currency and the common foreign and security policy, are struggling to find a realistic formulation ... The CAP is the culmination of over one hundred years of state support for agriculture in western Europe' (Ockenden and Franklin, 1995, pp. 1–2). The starting points and objectives of the Common Agricultural Policy were to stimulate agricultural productivity, to guarantee a profitable income to the people working in agriculture, to stimulate a balanced food market without surplus and shortage and to produce enough food at fair prices for the consumer. The CAP consists of a market and structural policy, which is characterized by three main principles: first, the existence of a common free trade market; second, common preference between the different member states, which means that products of member states take precedence over products of non-member states, and third, financial solidarity. The costs of the CAP are paid by all member states without taking into account differences of use between the member states.

Cork Declaration *(1996)*

The countryside across Europe and within national boundaries shows a great variety. This has a relevant impact on the rural policies, which attempt to promote the vitality and quality of rural areas. According to the Cork Declaration, policies must have a clear territorial dimension and provide the 'framework for self-sustaining private and community-based initiatives,' and they must be based on 'participation and a 'bottom up' approach, which harnesses the creativity and solidarity of rural communities.' National and European policies thus increasingly limit themselves to the formulation

of broad criteria, which offer guidelines for regional and local initiatives in close collaboration with the people involved (EU Council Regulation, 17.5.1999). Such a policy
shift can be clearly detected from a recent policy document issued by the Dutch Ministry of Agriculture (2000). It states that rural development is not a policy field that
should be developed by central authorities, but should be based on locally developed
plans that are initiated and implemented by the people themselves.

In short, rural development must be local and community-driven within a coherent European framework. Individual solutions must be found for each region in
the light of its inherent characteristics. In fact, the Cork Declaration can be characterized as a first policy statement that attempts to widen the scope of the CAP from a
purely agro-economic business approach to one that sees rurality within the context
of changing norms and values in society.

The Model of European Agriculture

The European Commission responded to the Cork Declaration by introducing the
so-called 'Model of European Agriculture' (MEA). This model supports the notion
that European farms are family businesses, integrated in a social community and in
local production conditions and market relations. Although sustainable production,
multifunctionality and the management of rural areas are taken as starting points, it
is rather unclear as to whether there is enough political willingness to change the CAP
in accordance with these principles. The formal function of the MEA is to serve as a
basis on which 'good agricultural practices' are defined. This concept should be applied to all agricultural areas in the EU. However, this standard should be translated
in order to fit local conditions. Services delivered by agriculture that go beyond the
standards of good agricultural practices should be additionally rewarded.

The MEA has received quite a lot of criticism. WTO partners regard it as a way of
defending agricultural subsidies. In addition, within the member states there is some
opposition, which in fact supports the criticism that some measures are simply 'win-
dow-dressing.' European agriculture must still prove its value to society. The willingness of society to support European agriculture has yet to be demonstrated, and it
is still unclear what the effects will be on world market competition. The question is
whether or not the current agreements in Agenda 2000 will meet these demands.

Agenda 2000

The MEA was elaborated on in the concept-Agenda 2000 (July 1997). This concept
was accepted by the European Council as a guideline for the development of European Agriculture in December 1997 (Luxembourg).

In the past few years, an increasing consciousness of consumers can be noticed in
aspects such as food security, environment, natural environment, welfare and rural
area development. These consumer concerns have been a major reason for reforming

the CAP. Of course, the Agenda 2000 was also drawn up against the background of the urgent need to change European agricultural production due to imminent overproduction, the new WTO round (particularly dairy policy), and the upcoming membership of countries in Central and Eastern Europe with large production potentials. A further reason is that the current CAP does not answer to the budget restraint of the EU (maximum expenditure of 1.27 percent of GDP of EU member states). Finally, the current milk quota system is the main restriction for structural changes and production growth in the European dairy sector.

In March 1999, the European Council (in Berlin) decided to accept the final Agenda 2000. This project has the following five formal *objectives*:

- Improving the competitiveness of EU agriculture both internally and externally.
- Guaranteeing and improving food safety and quality in line with consumer expectations.
- Ensuring a fair standard of living for the agricultural community and a stable farm income.
- Integrating environmental goals into the CAP.
- Promoting supplementary or alternative sources of employment and income in rural areas and thus contributing to the economic cohesion within the EU.

The most important *new policy instruments* introduced by the Agenda 2000 are:

- Cross-compliance: attachment of environmental conditions to the receipt of price-compensation support for agricultural products.
- A national envelope: direct subsidies paid in the form of a total sum related to the member state's share in the total production volume.
- Framework Regulation on Rural Development: Rural Development Plans (RDPs) will become obligatory for rural areas, and will be the basis for co-financing from Brussels.
- A new set-up for the structural funds: three new objectives replace the six former ones: (i) The poorest regions, which will receive highest priority, strict application, and two-thirds of the budget. This does not apply to the Netherlands. (ii) Areas with particular structural problems. In the Netherlands, this involves the northern region, and reconstruction and urban areas. (iii) Employment projects.

EU Management and Animal Diseases

During the past few years, the EU member states have been facing an almost constant series of epidemic livestock diseases. The most well known are mad cow disease (BSE), swine fever, and – during 2001 – the dramatic outbreak of Foot and Mouth Disease (FMD). However, food safety affairs related to, for example, salmonella and dioxin pollutions are expected to also have a huge (negative) impact on the economic future of the agricultural sector throughout the EU. Therefore, it is not surprising that the European Commission has taken initiatives to manage these continuing crises. During the management process, the European Commission has the role of a juggler. It has to anticipate, invent, organize, control, and sanction all these processes at the same time. Furthermore, these activities should be co-ordinated within the political frameworks conceived in common treaties and agreements under the leadership of the European Council (for example, the limit of expenditures to 1.27 percent of the EU-GDP, WTO agreements, Agenda 2000, 'schedule' of the EU enlargement, different domestic variables in member states – e.g. parliamentary and presidential elections).

At the same time, the European Commission seems to be quite uncomfortable in its attempts to legitimize its policies. This can be deduced from the language used in many EU documents. In particular, the formulation of instruments and measures demonstrates a highly technical rationality, which seems to be in great contrast to the 'thoughts and talks' *in* rurality. Of course, this makes it difficult to interpret the EU discourse incorporated in the official plans and instrumental statements delivered in a highly technical EU format, such as the 'seven-point EC-plan' to tackle the BSE crisis.[4] Fischler, the EU Commissioner for Agriculture, explained in the European Parliament (13.2.2001) that this 'seven-point-plan' aims to stop the ticking time-bomb caused by the following variables: growing stocks of meat, consumers' distrust, and increasing (transaction) costs of managing the BSE crisis. However, this 'seven-point-plan' is not a 'systemic revolution' because the European summit in Berlin (1999) had already decided to diminish the guaranteed prices and to impede extensive farming.

In order to interpret this technical policy framework as part of the EU discourse on agriculture, it is fruitful to consider some statements made by Franz Fischler in a recent interview, mainly concerning a new livestock disease crisis (at that time), i.e., the FMD (NRC, 7.4.2001):

- Each EU policy should be based on the opinions of scientists and experts, not on the emotions of farmers and other people.
- Vaccination cannot be used to rescue the animals. Vaccination of all European cattle today would make it impossible to definitively kill the FMD virus.
- It is easy to say that ethical criteria (not destroying healthy animals) should be on a politically higher level than export interests. However, for the individual farmer it does not make any difference whether he has to stop his firm because of the BSE and/or the FMD disease or because of the fact that no one wants to buy his meat.
- The EU is accountable also for the social well-being of the farmers: the EU buying-up regime has severely diminished the social-economic burden of the farmers.
- The EU reform of the Common Agricultural Policy includes, first, a more organic cultivation and a multifunctional enterprise by providing income-support to farmers and, secondly, the introduction of instruments (for example, Leader+) that sustain anti-migration in many European rural areas.
- The EU has made an important political step towards improving the image of European agriculture among consumers by establishing a Directorate for Consumer

Interests and Health as well as a Directorate for Agricultural Affairs. The first is involved in managing the livestock crisis, the second is involved in controlling disturbances in the (consumer) market.

- The European farm will be mainly family based in the future. However, the production process on the farm cannot be grounded in a romantic rural idyll, but should use all available technology and knowledge offered by science and industry.
- Intensive farming as such is not the problem. The real problems are: weak knowledge management throughout the agro-production chain (for example, information to the farmer about the quality of the food delivered to his livestock), insufficient control of manure pollution, too much transport of living livestock, marketing consumer demands, coordination between short-term crisis management and the agro-political reform agenda for the long term (Agenda 2000–2006).

In short, the dominant EU rural discourse is still based on a strong modernist belief in technical solutions delivered by an expert system. The EU policies are also extremely oriented to instrumental metaphors. A multi-functional, family-based enterprise will have a future in Europe, but the farmer will have to combine a (highly) technical attitude with a multifunctional approach in order to make his economic life less vulnerable. The EU approach to farmers seems to be extremely gendered as well. It seems that the EU policy-makers will not differentiate between male and female participation in farming, nor will they do so in the case of full-time and part-time involvement of the family partners in the farming business. Parallel (proto-) discourses as found in the Dutch situation (particularly the agri-ruralist and hedonist discourses, the multi-functional space discourse, the discourse concerning the establishment of new businesses and the wilderness discourse) seem to be outside the policy discourse at the EU level.

Of course, the EU has to cope with resistance as well. Therefore, organic farming will be more important than ever and the EC will promote it in good governance with the member states. Domestic obstacles will always be relevant, but not decisive during the long term. The system is still dominated by business-economic criteria. However, organic farming will never completely replace intensive farming. The modernist approach, although marketing will accommodate it to some consumer concerns, will continue to be the dominant agricultural production process. That is why the CAP has not passed a systemic revolution yet. On the other hand, the animal crises in combination with political agreements concerning world trade and enlargement affairs require serious reformations of the CAP to be implemented in the mid-term (2003–2004).

EU and Dutch Discourses and Practices of Rural Development in the Face of Power

It is assumed in this chapter that a power perspective can help reveal a plausible link between the EU policy discourse and the Dutch rural discourses and practices. Following Clegg (1989), Goverde and van Tatenhove (2000) power can be defined at three layers: i) 'power over' or episodical power; ii) 'power to' or dispositional power, and iii) '(shifts in) relations of power' or structural/systemic power.[5] Furthermore, in this section the EU policy discourse is perceived as a living context for the Dutch rural discourses and practices. The following gives a picture of Dutch ruralities in 2001. It thus includes many references to the management of the FMD crisis.

Power over/Episodical Power

With regard to the implementation of the CAP, the EU authorities have excessive power resources compared to the member states or to individual farmers. That is why the streets of Brussels are often the perfect public area for many farmers to demonstrate their anger as soon as they disagree with EU regulations. In fact, if farmers resist the national authorities, they are often encouraged to put the issue on the EU political agenda. In general, national autonomy has become rather proper and modest, particularly in policy fields such as agriculture, food security and rural development. This has often been illustrated by the public management during the various livestock disease crises in the last years.

These crises are often perceived by farmers and many consumers as the result of the failing agro-system in Europe, i.e., the CAP. Resistance is therefore oriented to the national (governmental and non-governmental) political leaders. An idiosyncratic example was given by the population and some farmers in a small town in the center of the Netherlands. This rural community expressed its local identity and mobilized its scarce power resources in both a symbolic and violent way by hanging dead swines in the trees and placing white small crucifixes along the roads for each animal killed. The protest eventually led to street fights and the erection of barricades. The inhabitants and their regional supporters were fighting against 'the (FMD) expert system,' which was being supported by special police forces and 'special slaughterers.' This case is remarkable, because on the one hand the dominant religious belief-system in this small rural community is very sensitive to public authority. People perceive public authorities as having God-given power and therefore political resistance is out of the question (i.e. dispositional power). On the other hand, this region has a tradition of resisting political centralization at the national as well as at the EU level. During the 1960s, the 'Farmers' party' – a rather populist movement of farmers and other small businesses – was founded here (Nooy, 1969).

In other places, rural communities (farmers, as well as others) have protested against the dominant utilitarian discourse supported by the national and supranational authorities. In particular, the anti-vaccination policy (to protect the export of meat in the – near – future) has lost much of its legitimacy. People have demanded a more 'humane' policy for the cattle. According to the agri-ruralist discourse, self-regulation and endogenous development were claimed as practical tools for rural communities to serve the needs of the urban people. Viewed from this perspective, it is remarkable that not all of the rural communities affected by the FMD crisis reacted in such a violent way. Some communities, particularly in the Frisian region, have shown deep solidarity and social cohesion among their inhabitants. This social capital was not used to fight against the power/ knowledge system of the public authorities, but to empower the victims of the crisis to make an economic rebirth. In a few other villages the cultural gap between native farmers and hobby (or urban) farmers diminished after the shared experience of losing cattle in the FMD crisis.

Dispositional Power

Only a few farmers were involved in the above-mentioned relatively violent public protest. The first acts of violence were even credited to a farmer of urban origin, labelled as an 'outsider.' It has been observed also that the streetfights were initiated mostly by 'established' local supporters of the farmers. However, many farmers were disposed to willingly accept the crude approach of the FMD expert system, i.e., to clear out farms as fast as possible, even for precautionary reasons. Of course, the EU and national rules in the official crisis-management scenario can explain this submissive behaviour. Any resistance by the farmers might reduce the transfer-payments related to this approach. Some farmers expressed publicly that they felt they were being forced to do so by coercive power. However, they also legitimated their dispositional behaviour by referring to the common interests of the sector.

Of course, the authorities – governmental and non-governmental at all levels of governance – argued similarly during the decades when the CAP regime was very profitable for the Dutch farmers and the Dutch economy. Nowadays, however, it is still uncertain whether the EU institutions as well as the legal 'powers in The Hague' effectively create the dispositional power necessary to guide the farmers into the regime of Agenda 2000 (including anticipation of a new WTO round and the EU enlargement). More empirical research seems to be urgent, for example, in order to discover whether 'local power-brokers' (opinion-leaders) function as real mediators between EU policies and the local agricultural entrepreneurs.

Structural Power

The CAP, the *acquis communautaire* (relevant for EU enlargement), the WTO regime, and the EU expert system form a systemic power that cannot be neglected, neither by any individual farmer nor by any (governmental or non-governmental) participant in an area-oriented reformation. Particularly during crises, it is obvious that state institutions should be prepared to use their monopoly of violence, i.e., to intervene with coercive power tools (persuasion by experts, inspectors, juridical procedures, police, army). It should be noted, however, that nation states are expected to activate their institutions of violence not only to guarantee national law and order, but also to implement the authority of supra-national and inter-governmental EU institutions. It can therefore be concluded that the power systems of the nation state and the EU are now strongly interwoven, even without a serious development of the 'second/third' pillar at the European level.

On the other hand, the system of structural power can be contested as well. This system produces and sustains inconveniences such as animal diseases, over-production, inefficiency by turning around huge amounts of money, bureaucracy, and inefficient production far above world market prices. However, the resistance to this systemic power has not been successful enough to either adapt the rules of the game (dispositional power) or to produce a systemic revolution. Until now, only a few relevant adaptations in distributive policy tools have been achieved at the level of episodical power. In accordance with some aims in the Cork Declaration some parts of the CAP budget will move to the structural funds. In general, however, the mainstream of agricultural production in the fifteen EU member states thanks to often successful lobbying by French and Italian (non-)governmental agricultural organizations and pressure groups (Keeler, 1987), will continue as a project dominated by a (neo-)modernization discourse. It is true that because the BSE crisis had a huge impact on the demand of meat, the BRD government is aiming to increase organic and ecological farming to up to 20 percent of total agricultural production in the year 2010. However, this implies that approximately 80 percent of the traditional, CAP-driven farming will be able to continue. Secondly, this assumes the *ceteris paribus* condition of all other relevant variables (WTO regime, EU enlargement).

Insofar as supra- and international regimes are mutually dependent on domestic political forces, it can be expected that the CAP regime will continue for the most part. Therefore, the context of Dutch rural development, as far as it concerns the EU level of governance, will still be dominated by policies that attempt to manage political and economic integration between anticipation (mostly of new international regimes) and resistance (mostly to different domestic variables) (Goverde, 2000b; Keeler, 1996). This was recently confirmed by the EU commissioner for agriculture, Franz Fischler (NRC, 7.4.2001): 'I don't want to change the rural areas of Europe into an agricultural museum. While the Dutch authorities want to fight against the agricultural crises with ethical ad-hoc politics, I believe we should use the technical capacities of science and industry as optimally as possible.' This statement seems to reflect the dominance of the neo-modernization discourse at the European level. Although the Cork Declaration and the MEA suggest the urgency of a more nuanced perspective, it is obvious that the other (proto-) discourses, as socially constructed in the Dutch policies and practices (see former section), have not yet found a legitimated counterpart in the EU political arena.

Constituent Power of Discourses in EU–Dutch Multi-Governance Polity

Constituent power of discourses in Dutch and EU rural policies can be defined as the impact (consequences: congruency or in-congruency) of discursive variety on the contents and the process of policy-making and policy implementation in the multi-governance setting of rural development. In this context the discourses concern mainly ideas, assumptions, goals, solutions, instruments, and conditions of policies in a certain field. This implies, first, that public institutions are involved in the production of these discourses and, secondly, that they include a 'normative leap' (Schön and Rein, 1994) implying the description of an 'undesirable situation' as well as the answer to the question concerning what should be done. It is assumed that the more (policy) discourses reflect the episodical ('power over'), dispositional ('power to') and systemic ('dominant power relations') dimensions of power, the more they constitute the policy that will be enhanced throughout the multi-level governance polity.

The constituent power of the EU and Dutch discourses is relatively high. Particularly the policy discourse at both levels of governance represents a rather clear and close congruency. A recent debate about the functioning and relevance of the CAP for the prosperity of the farmers and the Dutch economy confirmed this observation (*Internationale Spectator*, 2001, 2002). Of course, the EU represents an important systemic context for Dutch rural development policies and, indirectly, for the rural areas. Particularly in area-oriented policy processes, it is obvious that many agricultural and non-agricultural actors are interdependent. These participants are related in a loosely coupled system: a policy network. The rules of the game in these networks often dispose the participating actors to incrementally accept the power executed by national and EU policy frameworks. In general, it can be expected that participants in interactive policy processes will exploit fewer fixations and will perceive the future in the area in a more open way. New exchange relations can then develop and new creative solutions can be found. Of course, this configuration will promote new opportunities for governmental episodical power to emerge as well.

On the other hand, there is no complete congruency between the EU and Dutch rural polities. It is also obvious that a strategic perspective of the Dutch position within rural Europe will include threats as well as opportunities. For example, despite a high population density and a high degree of urbanization, there is still a contrast between urban and rural areas in the Netherlands. However, according to the EU (and OECD) definitions, there is hardly any rural surface area left in the Netherlands. Based on this conclusion, the Netherlands would no longer be able to claim subsidies from the CAP funds as far as regional support is concerned. Of course, this will not be accepted by Dutch policy-makers, who will argue that it is impossible to apply a uniform notion of rural areas to the whole of Europe. Another example of the lack of congruency can be found in the politics related to the CEE countries. The enlargement of the EU can be seen as a threat to Dutch claims on the CAP and structural funds budgets because the addition of large, non-urbanized rural areas may come at the expense of the already limited attention given to relatively urbanized rural areas such as in the Netherlands. Perhaps the Dutch authorities will feel coerced into finding allies within relatively urbanized rural regions throughout Europe. Regions that are similar in terms of population density, production and environmental pressure are, for example, the Paris Basin, central Belgium, South-East England, the area surrounding Milan, Nordrhein-Westfalen, the coastal belt Lisbon–Oporto, Hessen, and parts of Denmark. On the other hand, the enlargement can imply opportunities as well. Within Europe, Dutch agribusiness stands out as being capital intensive, export oriented and relatively small scale. This position can produce competitive advantages. Only 50 percent of the Dutch agri-production comes under the European market and pricing policy, whereas the sector has always predominantly consisted of independent farmers. Owing to these differences, it can be expected that the changeover from price supports to a liberalized market will go more smoothly in the Netherlands than in other countries and the development of rural areas will improve the cooperative tradition and the reform thereof.

Inclusion and Exclusion: Gender and the Constituent Power of Discourses

Because a discourse is relevant or 'true' in a specific context only, the power of a discourse has a limited state of constituency as well. That is why discourses have different functions related to a policy process: i) to give a signal of variety in perceptions, frames, and fixations to the political-administrative system; ii) to evaluate (support, facilitate, undermine) a specific policy; iii) to create a new (political) agenda or 'policy design'. Depending on its state of constituency as well as its function in a policy context, a discourse can empower the inclusion or exclusion of ideas, orientations, values, actors, goals, solutions, instruments, etc. For example, a discourse that highly empowers gender issues in rural policy can be relatively less constituent than other discourses in the enhanced rural policy and as such exclude male–female relationships in the range of solutions, for example, to generate a higher income for farmers or to give a multifunctional economic foundation to a certain rural area.

In the EU and Dutch policy discourses on ruralities and rural development little attention has been paid to the aspect of gender (see also the chapter by Bock and de Haan). While '. . . rural women take active part in practical initiatives, they do not figure in the process of rural policy decision making, neither do they make use of those policy instruments meant to stimulate bottom-up innovative initiatives' (Bock, 2002, p. 185). In the Dutch situation especially, the absence of women is rather surprising, because the Dutch government explicitly promotes the participation of rural inhabitants in the so-called 'processes of interactive policy-making'. According to Bock (2002, pp. 186–188) this can be explained by the fact that interactivity of the rural governance process is still very limited. The main actors are the government and established political organizations, which focus on economic interests and determine the selection of the political players. Furthermore, political culture is still dominated by disagreements and conflicting interests, while women tend to focus on practical solutions. Because policy-making and policy implementation are separated processes, women are much more involved (relatively) in the implementation phase; however, their engagement carries often high costs (expensive childcare, travel, immaterial costs of resistance, disapproval and pressure to group cohesion). This can also explain why male-female issues meet many obstacles in processes of policy-making. Finally, she refers to the high fence between the policy process and the initiatives of ordinary rural inhabitants (policy process and tools remain invisible and unknown to most rural people): policy instruments (subsidies) are not often used in women's initiatives and when women see that their proposals have very little chance of succeeding in the official circuits, they return to their original, gradual, and step-by-step-model of rural innovation.

Whereas the discourses mentioned above seem to neglect gender in rural development, these narratives are not gender-neutral. In fact, bringing politics back into debates about rural development demands an explicit inclusion of gender, class, and ethnicity. Nowadays, many residues of traditional neo-corporatist manners of governance – which traditionally supported the modernization project in agriculture – are obstacles for rural innovation in local politics. It can thus be expected that not only women, but also other newcomers in the rural areas (settlers from urban origin, migrants) are not incorporated into processes of rural policy-making, although they have participated in initiatives of rural innovation. This phenomenon implies already that although policy discourses at the EU and the Dutch level show some congruency, the policy-makers and the rural population can exclude each other and as such they can live in two different worlds.

The explanation of politics about ruralities and in ruralities requires, therefore, in addition to discourse analysis an in-depth analysis of the mutual (horizontal and vertical) dependencies between the actors involved in rural policy-making. Of course, it is conceivable that the consistency between discourses and mutual dependency of actors will be reflected in the (episodical, dispositional and systemic) power construction of the rural policy. However, at this point it should also be recognized that actors have different access to policy resources (formal and informal), that they participate in different patterns of social and political cohesion and that they are disposed by different cultural attitudes, codes, rituals, and perceptions of the desirable or undesirable situation. In short, although actors are involved, even mutually dependent, in the same multi-level governance polity, they often define and approach the 'public space' (what is and has to be deliberately debated) of this polity differently.

Conclusion

From this section we can conclude that discourse analysis often suggests a large discursive variety in everyday life. Because the leadership in policy processes focuses often on the rationality of strategies, goals and means, the discursive diversity – particularly of 'everyday discourses' – is furthermore underexposed. How can this be accounted for? Many issues 'about' rurality and 'in' rurality are complex, i.e., they are a mixture of structured or non-structured problems and solutions (strategies, options, and management styles), which should follow a certain sequence but, in fact, are pro-

duced in a parallel way and then cause much ambiguity and uncertainty. Furthermore, regional institutional embeddedness is often incongruent with the life-circle
of many national policy regimes and their technical rationality. Last but not least, the
participation of citizens is often indirect, fragmented and only incidental, whereas
the involvement of governmental institutions and organizations of civil society is
permanent. In short, it is urgent that extra attention be paid to the differences between the 'world of citizens' and the 'world of policy makers.'

Discussion

While in theory the constituent power of EU and Dutch (policy) discourses 'about'
and 'in' rurality can help produce a consistent and coherent multi-level governance
polity, in practice its judgment depends on the standing of the actor. Although the official EU agribusiness and rural development discourse reflected in rural policy documents, and the Dutch domestic rural discourses seem to have developed in a similar
way, the congruency in these rural policy discourses is mostly symbolic. Particularly,
when the long-term policies (CAP and EU enlargement) as well as the policy processes
during short-term crisis management around animal diseases are put into practice,
the divergences emerge into the public arena rather rapidly. Two opposing processes
can help explain this discrepancy. On the one hand, the EU has to play an extremely
complex role as political mediator between global institutions (WTO, FAO) and domestic governance scales in the member states. On the other hand, the EU as a supra-
national agro-political institution is a dominant actor in a power balance game between itself and many member states with often opposing domestic rural political organizations, particularly during crisis management of agricultural and rural affairs.

From the perspective of the Dutch actors in rural development, it is clear that the
EU is an important, although not always easily accessible and sometimes even weak,
democratic institution that functions *de facto* as the fourth level of government. This
implies that lobbyists should be very selective in their orientations to the policy- and
decision-makers. In this way, the governance system has become extremely fragmented. This has produced different opportunities, and even larger inequalities, to
influence the relevant governmental actors. Although this 'coin' can have a positive
as well as a negative side, it is obvious that the behaviour of the official functionaries
in governmental institutions has become less predictable. National authorities in the
Netherlands, who have formally accepted the change in the policy paradigm included in the Cork Declaration and the MEA as well as demanded by domestic regional
and local actors, enhance this situation also. However, they could not realize a fitting
set of policy tools, or an upgrading of the organizational institutions (Bekke and de
Vries, 2001, p. 187). As a consequence, there has been a loss of legitimacy. The decline
of trust in government is apparently not easily restored, neither by new rules nor by
new interactive procedures. In fact, these intermediary policy processes have often

produced new problems of managerial effectiveness and efficiency as well as political representation and accountability.

Since the early 1990s (i.e., the debate about the Maastricht Treaty and the Mac-Sharry Reform), the main policy framework in the European Union concerning agriculture has been challenged by many new public and private actors. Besides the traditional key actors, farmers and their interest groups as well as politicians and policy-makers, new actors have entered the political, social, economic and cultural arenas of rural areas. These actors have produced a new rural fabric, which has become extremely complex. As Vihinen (2001) has argued, a new contract between agriculture and society will not only include agreements on traditional products (milk, cereals, beef) and production factors (land, labour, capital, knowledge), but will also concern consumption items such as landscape, nature, environment, rural development, food safety and, at a more abstract level, space, biosystems, animal welfare, GMO, production methods, ethics and morality.

Notes

1. De Bruin (1997) has distinguished between conservatives (who hang on to the project of modernization), neo-modernists (who adapt the modernization project with optimal technical methods), and pluralists (who construct a counter-movement; more or less parallel to Frouws' agri-ruralist discourse and that of van der Ploeg, see below). Hidding et al. (1998) describe five narratives concerning urban–rural relations in the sphere of spatial planning: city and countryside are seen as anti-poles, as networks of activities, as ecosystems, as configurations of places, and as objects of property. Van der Ploeg (2000) has developed a diachronic perspective on the discursive variety: rural development is deeply involved in rearranging the dominant paradigm (policy, practice and theory) of modernity.

2. The methodological background of this argument is inspired by Habermas (communicative acting: stories of people should be studied not only as 'text' but also 'in context': do people do what they say?), by Foucault (1980: the relation of discourses and social practices and power, particularly the dominant power of expert-systems), and by Hajer (1995: a discourse is dominant if it is based both on discourse institutionalization and -structuration). Discourse institutionalization concerns the translation of a discourse into institutional arrangements, for example policy programmes, the constitution of a new advisory board, or a new form of public–private cooperation. Discourse structuration reflects the idea that the credibility of actors in a specific domain urges them to use representations of that discourse.

3. Throughout the EU member states as well as in the other Nordic countries and Switzerland the total acreage used for organic farming increased (in the period 1998–2000) by 55.6 percent. In the Netherlands this figure was 31.4 percent. Although the highest number of organic farms can be found in Italy (43,698), the highest percentages of organic farming (except Liechtenstein: 16 percent) are in Austria (8.9 percent), Switzerland (6.8 percent), Finland (6.1 percent) and Denmark (5.2 percent). The Netherlands (1,216 organic farms) is still on a relatively low level (1.2 percent). This figure corresponds with that of the largest EU countries: France (1 percent), Germany (2 percent), Italy (1.8 percent), and the UK (0.7 percent).

4. Seven-point anti-BSE plan (translated from NRC, 14.2.2001): i) Impede 'organic farming' by restricting permission to use fallow land for green cattle food (particularly clover). ii) Reduce premiums for male cattle and meat cows from 2/ha to 1.8/ha. iii) Introduce special premiums for male cattle up to a maximum of 90 heads/firm. iv) Introduce new premium rights for male cattle that will be delivered to individual firms in order to stop the balancing between intensive and extensive farming. v) Stop the 'buy-up for destruction' of all cattle above 30 months of age, because – since 1.4.2001 – all countries have to implement the BSE test. Then start a new buying-up regime (based on the usual 70 percent EU/30 percent member-state rate): member states can then stock meat on their own costs or take it from the market. Non-commercial delivery of this meat to people in crisis-situations is possible. vi) Limit premiums for cows which are used for breeding meat-calves by the introduction of the criterion that 40 percent of the herd should be heifers. vii) Discontinue maximum of 350,000 tons in the intervention buying-up arrangement for the years 2001 and 2002.

5. These three layers in the concept of power were described in the introduction to part I.

References

Alvesson, M. and K. Sköldberg (2000), *Reflexive Methodology. New Vistas for Qualitative Research*. London, Sage.

Bekke, H. and J. de Vries (2001), *De ontpoldering van de Nederlandse Landbouw*. Apeldoorn, Garant.

Boonstra, F. (2002), Leren van plattelandsdiscoursen: omgaan met verscheidenheid in de regionale beleidsarena. *TSL* 16 (4), pp. 242–257.

Bruin, R. de (1997), *Dynamiek en duurzaamheid. Beschouwingen over bedrijfsstijlen, bestuur en beleid*. PhD-thesis Wageningen University, Wageningen.

Buddingh, H. and R. Moerland, Interview with Franz Fischler. NRC 7.4.2001, 35, Z3.

Clegg, S.R. (1989), *The Frameworks of Power*. London, Sage.

Council for the Rural Areas (2000), *European Integration and regional diversity: a challenge for the Dutch Ministry of Agriculture*, (publication 00/1a). Amersfoort.

European Commission (1997), *Agenda 2000: voor een sterke en uitgebreide Unie*. Brussels

European Conference on Rural Development (1996), *The Cork Declaration, A Living Countryside*. Cork.

Foucault, M. (1980/1971), *L'Ordre du discourse*. Paris, Gallimard.

Frouws, J. (1998), The contested redefinition of the countryside; an analysis of rural discourses in The Netherlands. *Sociologica Ruralis* 38 (1), pp. 54–68.

Goverde, H. and H.J. de Haan (2002), The politics of rural development in The Netherlands. In K. Halfacree, I. Kovach and R. Woodward (eds), *Leadership and Local Power in Contemporary Rural Europe*. Aldershot, Ashgate, pp. 33–58.

Goverde, H. (2001), Managing integration and marginalization for the New Europe. In H. Jussila et al. (eds), *Globalization and Marginality in Geographical Space. Political, Economic and Social Issues of Development in The New Millenium*. Aldershot, Ashgate, pp. 61–76.

Goverde, H. and J. van Tatenhove (2000), Power and policy networks. In H. Goverde et al. (eds), *Power in Contemporary Politics*. London, Sage, pp. 96–111.

Goverde, H. (2000), EU and agro-politics. Managing integration between anticipation and resistance. In H. Goverde (ed.), *Global and European Polity?* Aldershot, Ashgate, pp. 253–268.

Hajer, M. (1995), *The Politics of Environmental Discourse. Ecological Modernization and the Policy Process.* Oxford, Oxford University Press.

Hidding, M.C. and J. Westerhof (1999), Op zoek naar een nieuw verhaal over stad en land: planning in discoursperspectief. *Planologische diskussiebijdragen*, pp. 137–146.

Internationale Spectator (2001, 2002), Debate about the future of agriculture. 2001, pp. 440–446, 615–621, and 2002, pp. 55–57.

Keeler, J.T.S. (1987), *The Politics of Neo-corporatism in France.* New York, Oxford University Press.

Keeler, J.T.S. (1996), Agricultural power in the European Community: Explaining the fate of the CAP and GATT negotiations. *Comparative Politics*, January, pp.127–149.

Ministry of Agriculture (2000), *Nota Voedsel and Groen. Het Nederlandse agro-foodcomplex in perspectief.* Den Haag.

Nooy, A.T.J. (1969), *De Boerenpartij. Desoriëntatie en radikalisme onder de boeren.* Meppel, Boom.

Ockenden, Jonathan and Michael Franklin (eds) (1995), *European Agriculture. Making the CAP Fit the Future.* London, Chatham House Papers.

Ploeg, J. D. van der (1999), *De virtuele boer.* Assen, Van Gorcum.

Ploeg, J.D. van der et al. (2000), Rural development: from practices and policies towards theory. *Sociologia Ruralis* 40 (4), pp. 391–408.

Raad voor het Landelijk Gebied (2001), *Kansen voor de biologische landbouw. Advies over de kansen voor biologische landbouw in Nederland in de periode tot 2015.* Amersfoort, publicatie RLG 01/3.

Schön, D.A. and M. Rein (1994), *Frame reflection: Toward the Resolution of Intractable Policy Controversies.* New York, Basic Books.

Vihinen, H. (2001), *Recognising Choice. A Study of the Changing Politics of the Common Agricultural Policy through an Analysis of the MacSharry Reform Debate in Ireland and the Netherlands.* Helsinki, Agrofood Research Finland, Economic Research (MTTL), publication no 99.

PART II

GENDER AND RURAL DEVELOPMENT IN ACADEMIC DISCOURSE

Chapter 8

Introduction to Part II: Gender and Rural Development in Academic Discourse

Mireia Baylina and Bettina Bock

Gender is one of the primary elements of social structure, and of power relations in particular (Faulks, 1999). Therefore, it can be expected that changing local power relations in rural development have affected gender relations as well (Shortall, 1999). Part II focuses on the specific relation between gender and power in rural development. It describes fundamental changes in gender relations in European rural areas during the last three or four decades. While the chapters in Part I are based on empirical research, Part II focuses on the study of rural gender relations as an evolving discourse in the context of rural development processes and the development of the social sciences in different European countries. The object of analysis is not rural gender relations *per se*, but how rural gender relations have been studied, interpreted and explained by academics, who themselves became embedded incrementally in the practice, the policy and the narrative of rural development. The changing gender relations are thus revealed only indirectly.

This part of the book has three purposes. First, it contributes to our understanding of how changing socio-economic and political (rural) contexts affect the position of (rural) men and women. For example, this position has been influenced, indirectly, by the redistribution of resources and redefinition of rules and regulations. Of course, the changing perception, interpretation and evaluation of their position in and access to rural politics and rural development have produced changes in rural gender relations as well. To give just one example: Whereas rural women were not expected to be engaged in politics and even punished for doing so (by gossip, for instance) until about ten years ago, most European countries have now accepted 'equal opportunity' policies in order to promote rural women's participation in rural politics and policymaking. Consequently, this re-interpretation of rural women's 'duties' and 'capacities' creates new opportunities in rural development. However, it constructs new frames about women's political 'non-participation' and 'backwardness' at the same time. The more new opportunities for participation have been incorporated in the formal interpretation of rural politics and its relevant political institutions, the more informal political activities and structures of rural women seem to become 'invisible' and 'devalued' (Bock, 2002).

The second purpose is to show how understanding the complexity and the ambiguity of gendered processes in rural areas enables us to contribute to the development of effective gender-sensitive policies concerning rural innovation.

Finally, the country-based reports in Part II aim to reflect the theoretical development and growth of rural gender studies in the last decades. The shift in research emphasis from establishing rural women's invisibility and disadvantage in a world defined by male farmers towards an analysis of the different ways women and men themselves experience and define rural life is a very relevant and significant theoretical development (Whatmore, 1994). Particularly, it shifts the focus of attention from women's disadvantages and disability in adopting male behaviour towards women's agency and the rationality of their different behaviour (Bock, 2002; O'Hara, 1998). To understand how gender colours rural life, it is important to explain the gender-specificity of other basic concepts like rurality, development and innovation. This again is crucial for the development of effective rural development policies.

The close relationship between rural power and decision-making and gender relations is revealed in governance. Through an analysis of women's representation and initiatives in the political process, and policy outcomes, the gendered nature of power, policy processes and policies in rural communities is undeniable. Unfortunately, however, there has been little research until now on the role of gender relations in the rural policy-making process (Little, 2002), although the emerging interest in new rural governance by rural researchers provides an opportunity to begin to interrogate the gendered nature of the distribution and outcome of political power in rural areas.

The significance of gender in rural power relations becomes obvious when rural governance is analysed in terms of men's and women's participation in decision-making bodies and the representation of their (different) interests, problems and priorities in rural development policies and programmes. Research in Northern Ireland (Shortall, 2002), the United Kingdom (Little, 2002) and the Netherlands (Bock, 2002) has revealed that women are neither quantitatively nor qualitatively well represented in rural politics. As a result, the legitimacy and accountability of rural governance may be questioned as well (Wiskerke et al., 2002). Although the new interactive and bottom-up governance approaches are supposed to enhance rural democracy and support the participation of rural inhabitants, there are serious indications that new and old elites reclaim their powerful position and recreate situations of inequality and exclusion (Shucksmith 2000; Shortall, 1994; Lowe et al., 1994.) There is, thus, an urgent need for more and more comparative research on the genderedness of new rural governance modes and arenas.

Academic discourses on rural gender relations have been developed in close relationship with rural everyday life and rural development practices. Motivated by the women's movement, (female) rural scientists discovered rural women as an interesting field of study. By introducing a feminist perspective they made women visible, 'uncovered' the inequality and injustice of their social and political position and increased rural women's awareness of their situation. This in turn has contributed to the empowerment of rural women and their capacity to claim a more powerful position on the farm, in farm organizations as well as in rural society as a whole. As a result not only the self-esteem and agency (or social capital) of individual rural women increased: the collective social capital of rural women's organizations expanded as well (Townsend et al., 1999; Léon, 1997). However, this individual and collective empowerment is limited. Rural women certainly have more opportunities to realize their own projects and ambitions than they had before, and rural women's organizations seems to have gained more access to decision-making institutions, but rural gender relations are still far from equal in most European countries. For the most part, rural women's power depends largely on the amount of individual and collective social capital they have. However, constrains originate also in part from rural women's multiple and conflicting interests and ambitions. Farm women as well as farm women's organizations often suffer from the dilemma of being (or representing) a farmer and a woman at the same time. When women push their interests in rural development policy, it may damage the position of the household as farmers or contradict specific projects of their husbands. Therefore, rural women's involvement in rural governance processes requires not only support and empowerment; it also demands a redesign and reorganization of rural policy-making.

The chapters presented in Part II analyse the changing academic discourse of rural gender relations in four European and one Southern Mediterranean country: The United Kingdom, the Netherlands, Norway, Spain and Tunisia. The selection may seem somewhat surprising at first glance. However, it is especially interesting as it offers the opportunity to compare different conditions concerning the type of agriculture, degree of rurality, type of gender relations and the main objectives of rural development policy. The UK is an urbanized society with remote rural areas. Norway is considered to be one of the trendsetters with respect to gender equality and it has very remote, isolated rural areas. The Netherlands has highly industrialized agriculture and consists of a mainly urbanized countryside. Spain has traditionally rather conservative gender relations at the local and family level and has many very remote rural regions. Finally, Tunisia offers a contrasting picture in nearly all respects. It is a country that demonstrates the importance of integrating gender in development policy in order to resolve the enormous problems of poverty and famine.

These five chapters widely document women's role in rural society and its evolution over the last thirty years. All chapters explicitly link the changing socio-economic and political context with a changing perspective in rural gender studies. They describe which issues have been studied and how the theoretical and methodological approaches have changed. From this analysis it may be deduced, first, how complex the concept of 'rurality' is and how its significance depends on the social and environmental context. Secondly, the chapters point to the divergent historical paths into modernity that European and Southern Mediterranean regions have followed.

Thirdly, they reveal the relevance of rural women's contribution to the rural restructuring process, despite its neglect in the development of plans and policies. Finally, rural feminist theory has contributed to the deconstruction of important concepts such as work, politics and rural households. It has uncovered new issues of interest such as the perception of rurality, rural otherness, quality of life and sexuality. But it has also broadened the development of other current issues such as new rural governance or innovation by pointing out how important the gender-specific study of such issues is.

References

Bock, B.B. (2002), *Tegelijkertijd en tussendoor. Gender, plattelandsontwikkeling en interactief beleid Wageningen* (Ph.D. thesis).

Faulks, K. (1999), *Political Sociology. A critical introduction*. Edinburgh, Edinburgh University Press.

León, M. (ed.) (1997), *Poder y empoderamiento de las mujeres*. Santa Fe de Bogotá, TM Ediciones.

Townsend, J. et al. (1999), *Women and power*. London, Zed Books.

Little, J. (2002), *Gender and Rural Geography. Identity, Sexuality and Power in the Countryside*. Harlow, Prentice Hall.

Lowe, P. et al. (1998), *Participation in rural development: a review of European experience*. Newcastle upon Tyne, Centre for Rural Economy.

O'Hara, P. (1998), *Partners in Production; Women, Farm and Family In Ireland*. New York/ Oxford, Berghahn Books.

Shortall, S. (2002), Gendered agricultural and rural restructuring: a case-study of Northern Ireland. *Sociologia Ruralis* 42 (2) pp. 160–176.

Shortall, S. (1999), *Women and Farming; Property and Power*. London/New York, Macmillan.

Shortall, S. (1994), The Irish rural development paradigm – an explanatory analysis. *Economic and Social Review* 25 (3) pp. 233–260.

Shucksmith, M. (2000), Endogenous development, social capital and social inclusion; perspectives from LEADER in the UK. *Sociologia Ruralis* 40 (2) pp. 208–218.

Whatmore, S. (1994), Theoretical achievements and challenges in European rural gender studies. In L. Van der Plas and M. Fonte (eds), *Rural Gender Studies in Europe*. Assen, Van Gorcum, pp. 39–49.

Wiskerke, H., B.B. Bock, M. Stuiver, H. Renting (2002), Environmental co-operatives as a new mode of rural governance. *Netherlands Journal of Agricultural Science* 50 (3).

Chapter 9

Discourses on Gender and Rural Restructuring in the United Kingdom

Rachel Woodward

The aim of this chapter is to review the main features of rural restructuring and its relationship to gender issues within the United Kingdom (England, Scotland, Wales and Northern Ireland). I will note both basic socio-economic and cultural trends and issues in the relationships between gender and rural restructuring in the UK, and document the different conceptual and theoretical approaches taken towards the study of this relationship. This overview will look at the academic context within which these issues have been studied and will then explore key issues in the gender/rural restructuring relationship. The coverage provided is indicative, rather than exhaustive.

The Context of Gender Studies in the UK

Feminist social science has had two distinct impacts on British social science as a whole over the past thirty years. The first impact has been in expanding the legitimate territory for inquiry for many disciplines (sociology, geography, anthropology, political science) to include the study of feminist and women's issues. The development of feminist social science from the 1970s in the UK started with assessments of women's socio-economic position and patterns of gender inequality, including the examination of the relationships between paid and unpaid work, the politics of the family, social welfare issues and health matters. Gender studies during the 1970s and 1980s generally concentrated on the investigation of women's inequalities relative to men, and the study of these inequalities is now generally accepted as a legitimate area for social scientific inquiry.

In the past decade, the scope of gender studies has expanded to consider issues of consumption, identity, representation, power and the body, as a direct consequence of the second impact of feminist scholarship within the social sciences in the UK, which has been the influence of feminist theory. This development is not unique to the UK, of course; developments in feminist political economy and post-structuralist traditions have been crucial across the globe. Such developments include conceptual work on the body and space, and theories of the social construction of gender identities and representations, their causes and consequences. The validity of studies of masculinity as well as femininity as social constructions has contributed to the notion of gender studies rather than just women's studies. Gender studies is now generally interpreted as the study of both men and women and the influences of gender on all forms of social life.

In the UK, much feminist social science is conducted within academic departments with distinct disciplinary affiliations. For example, there is an extremely strong tradition of feminist sociology, and the vast majority of sociology departments in UK universities will have individuals or groups conducting research on gender and feminist issues. Academic research networks have been influential in establishing the disciplinary credibility of feminist studies within specific disciplines – see for example, the work of the Women and Geography Study Group of the Royal Geographical Society and Institute of British Geographers. In addition, gender studies are also conducted under the rubric of Women's Studies or Gender Studies, in multi-disciplinary research environments. Women's Studies and Gender Studies have become established academic disciplines recognized for research, study and (importantly) funding. Such multidisciplinary centres range from autonomous, funded research and teaching departments, to smaller research networks; the Universities of Newcastle, Leeds, York and Exeter, for example, all have centres of varying sizes for gender study.

The study of women and rural restructuring has emerged as a consequence of the expansion of feminist studies and feminist theories, and the majority of those currently writing on these issues in an academic context will have been influenced by the emergence of feminist scholarship in the social sciences since the 1970s. Indeed, some of those writings have themselves contributed significantly to the broader process by which feminist social science has become established and respected – within geography, for example, the work of Little and Whatmore has been very influential. The broad disciplines from which gender and rural restructuring have been studied include rural sociology, human geography, anthropology and political science. Different disciplines have (inevitably) produced different foci for study, although there is of course a blurring of disciplinary boundaries. Significantly, there has been a relative lack of comparative European research on this specific issue of gender and rural restructuring, relative to the coverage in the UK literature of the Anglophone developed world. Although there have been the inevitable complaints about the marginalization of gender issues within rural studies (Whatmore et al., 1994), as Little (2001) demonstrates, a rich literature on gender and rural issues has gradually emerged over the past two decades.

Key Themes and Issues in the Study of Gender and Rural Restructuring in the UK

This section considers the four key themes or issues in the study of gender and rural restructuring, these being studies of economic life, studies of service provision, studies of rural development policies and studies of the cultural construction of rurality. This overview is not exhaustive in its coverage of issue, but rather focuses on those issues which have driven forward some broader theoretical debates in rural studies. This chapter concentrates on socio-economic and cultural issues; other themes, such as the study of women and environmentalism (see, for example, Buckingham-Hatfield, 1999) have had less impact on rural restructuring debates and so are not covered here.

Economic Life

Predictably, focus for much early work on economic issues looked at women in the agricultural sector. Feminist studies have made an important contribution to debates on the role and conceptualization of domestic labour, in particular by highlighting the unique position of the family farm and the centrality of women's work in the household to the operation of the farm business (Little, 1997; Whatmore, 1991). Gasson (1992) examines the considerable contribution made by women to farming households. Much of the literature on women in farming households is usefully summarized by Whatmore (1993), who examines the study of issues such as the crisis in reproduction in family farms and the contribution this has made towards farm women's search for greater autonomy and opportunity. The exodus of women from farming has been interpreted by some as a response to the archaic gender relations associated with the family (Symes, 1991, quoted in Whatmore, 1993). Also noted has been the role of farm diversification in increasing women's autonomy *in situ*, and securing the future of the family farm (Deseran and Simpkins, 1991, quoted in Whatmore, 1993). The role of agricultural training has also been considered (Shortall, 1996). Shortall argues that agricultural training curricula privilege the work most frequently done by men; agricultural training can be viewed as a source of information and status. It is also an arena in which the gendering of agriculture is produced and reproduced.

For a broader examination of economic issues such as basic rural employment patterns for women, see Braithwaite (1994), Little et al. (1991), and for a theoretical treatment of this issues, see Little (1991). Elsewhere, Little (1997a) examines the constraints on women's paid employment in rural labour markets, arguing that women's economic roles within the rural household and community are both bound up with a set of myths, assumptions and expectations about rural society and culture (Little, 1997a). Changing configurations of work and family life have also been considered as part of a project assessing the situation for women in the context of labour market changes coupled with shifts in expectations about gender roles and patterns of family life (see Mauthner, McKee and Strell, 1999).

Services

A regrettable feature of rural restructuring in the UK over past thirty years has been a decline in levels of service provision in rural areas. The associated difficulties in ac-

cess to health, education, retail, leisure and legal and associated services and transport have been a focus for rural studies. Service provision includes both supply of
such services as well as access issues. Research suggests that women have been hit
disproportionately hard by such declines, particularly younger women, lone parents
and older women. Stone (1990) examines the problems associated with the lack of
childcare for women in rural areas, particularly of pre-school age children (compul-
sory education for children starts at the age of five). Barker (1997) has looked at af-
ter-school child care provision in rural areas in terms of urban–rural differences in
type and levels of provision. Studies of poverty and deprivation in England and Wales
have included assessments of service decline as a contributing factor, but have tended
to include women as a sub-set of those defined as deprived (Cloke et al., 1994, 1996).
More recent work assessing the quantification of levels of poverty and social exclusion again has an implicit rather than explicit gender dimension (Dunn et al., 1999).

The structure of the rural labour market in the UK is such that women are often
trapped in low-paid, seasonal or insecure employment. Chapman et al.'s (1998) study
of poverty and exclusion in rural Britain shows differences between men and women
in employment levels and low pay; unemployed women find it harder to move into
higher-paid employment in rural areas than unemployed men. There has been little
study of the specific causes and consequences of deprivation and social exclusion for
woman. Research looking specifically at rural women in Scotland points to the need
for gender strategies in rural policy (Clark, 1997).

Policy and Politics

Rural areas in the UK have been the recipient of much European Union finance
for rural development, under both Structural Funds (Objectives 1 and 5b, the new
Objective 2) and Community Initiatives (LEADER, RECHAR). However, there has to
date been little research looking explicitly at the impact of such policies on women
(although see Braithwaite, forthcoming). In Scotland, the lack of research on rural
women's inclusion and perspectives in the policy-making arena has also been noted
(Breitenbach, 1999), particularly in decision making at all political levels, and in the
public and private sectors, although women's participation has always been higher
in the voluntary sector (Myers, 1999). This latter point echoes Little's findings in her
examination of rural women's voluntary work and its role in the empowerment of
rural women. Voluntary work can be both an opportunity for women or an exploitation of their traditional role; what is clear is that both consequences reflect the ways
in which rural women's voluntary work supports a particular construction of rurality (Little, 1997b). In general, social and community life in the countryside is dominated by activities undertaken by women (Little and Austin, 1996). Women are more
likely to predominate in unpaid and voluntary positions within community organizations such as the church, which provide an important focus for rural communities
(Seymour and Short, 1994).

Little and Jones (2000) bemoan the lack of appreciation of the role of gender identities within studies of rural community politics and governance. They go on to look at masculinity within the analysis of policy processes in rural areas, looking not just at outcomes of the process, but also at gender (specifically constructions of masculinity) in the formulation of policy and the development and operation of the policy process. Beyond this, the new rural governance literature has been rather gender blind; this literature looks at wider political issues and political organization within the context of rural restructuring. Local political networks exist and have been the subject of study – see, for example, Shortall's (1994) examination of farm women's groups and local capacity building. However, the gender dimension has been lacking from studies of rural governance, with some notable exceptions. For example, in Northern Ireland the increasing sensitivity of the state to gender issues within the policy process has been noted by Shortall (1999) in her study of CAP reform and new equality legislation put in place by the Belfast Agreement 1998 and Northern Ireland Act 1999. This legislation safeguards rights and obligations to promote equality of opportunity throughout Northern Ireland. A consequence has been that the Department of Agriculture of Northern Ireland is now seen as more willing to consider and act on gender equality issues in farming and rural development than had been the case in the past. Shortall argues that the political situation has advanced position of women in farming to a greater extent than the restructuring of agriculture.

Rurality and Gender

The influence of developments in social theory is reflected in the growing number of studies of rurality as a social construct and its relationship with gender. Philo's influential paper on neglected rural geographies (Philo, 1992) included women as a rural 'others,' conceptualized as those often excluded from consideration within rural studies. Further developments looking at the ways in which different discourses of rurality operate have noted a gendered dimension. See for example Hughes' (1997a) examination of the relationships between social constructions of rurality and femininity. Hughes (1997b) has also examined women's experiences of community and the ways in which this is shaped by how women both adhere to and contest dominant representations of gender identity within rural discourses; participation in community can be a duty rather than a joy. Jo Little's work on women and voluntary work (above) is also located in a theoretical framework that takes a dominant discourse of rurality (the rural idyll) as its starting point. The relationships between rurality and sexuality have also been assessed within this theoretical framework (Bell and Valentine, 1995).

Qualitative methodologies have dominated the social constructionist approach to the study of discourses of rurality, particularly ethnographic approaches. Such interpretative approaches draw on discourse analysis, chosen as the most appropriate methodological tools to investigate the ways in which social constructions (of gender identities and of rurality) operate.

The gendering of rurality has also been examined by Jones (1999), who has looked at the associations drawn between male children and nature. Jones argues that the 'natural' gender of childhood is constructed as male (Jones, 1999). The gendering of rurality is present in a variety of locations (see special issue of *Rural Sociology* on rural masculinities). My own work in this area has been on the connections between masculinity, rurality and the military (Woodward, 1998, 2000).

Rurality, gender identity and national identity have also been linked. Jones has examined the ways in which Welsh women link a sense of Welsh identity to senses of community in which they as women are defined roles (Jones, 1997).

References

Barker, J. (1997) *Rural out of School Child Care Provision: A Profile*. London, Kids Club Network.

Bell, D. and Valentine, G. (1995), Queer country: rural lesbian and gay lives. *Journal of Rural Studies* 11 (2), pp. 113–122.

Braithwaite, K. (forthcoming), *The Impact of Women's Networks on Rural Development in Ireland, Sweden and Cumbria, England*. PhD thesis in preparation.

Braithwaite, M. (1994), *The Economic Role and Situation of Women in Rural Areas*. Luxembourg, Green Europe European Commission.

Breitenbach, E. (1999), *Research on Women's Issues in Scotland: An Overview*. Women's Issues Research Findings No. 1. Edinburgh, Scottish Office.

Buckingham-Hatfield, S. (1999), Gendering Agenda 21: Women's involvement in setting the environmental agenda. *Journal of Environmental Policy and Planning* 1, pp. 121–132.

Chapman, P. et al. (1998), *Poverty and Exclusion in Rural Britain: The Dynamics of Low Income and Employment*. York, Joseph Rowntree Foundation.

Clark, G. (1997), *Rural Women: Gender Relations and Socio-Economic Change*. Rural Scotland Today Policy Briefing. Perth, Rural Forum.

Deseran, T.A. and Simpkins, N. (1991), Women's off-farm work and gender stratification. *Journal of Rural Studies* 7, pp. 91–98.

Gasson, R. (1992), Farm wives: their contribution to the farm business. *Journal of Agricultural Economics* 43 (1), pp 74–84.

Hughes, A. (1997a), Rurality and 'cultures of womanhood': Domestic identities and moral order in village life. In Cloke P. and Little J. (eds), *Contested Countryside Cultures: Otherness, Marginalization and Rurality*. London, Routledge, pp. 123–137.

Hughes, A. (1997b), Women and rurality: gendered experiences of 'community' in village life. In Milbourne, P. (ed.), *Revealing Rural 'Others': Representation, Power and Identity in the British Countryside*. London, Pinter, pp. 167–188.

Little, J. (1991), Theoretical issues of women's non-agricultural employment in rural areas, with illustrations from the UK. *Journal of Rural Studies* 7, pp. 99–105.

Little, J., Ross, K. and Collins, I. (1991), *Women and Employment in Rural Areas*. Research Report No.10. London, Rural Development Commission.

Little, J. and Austin, P. (1996), Women and the rural idyll. *Journal of Rural Studies* 12, pp. 101–111.

Little, J. (1997a), Employment marginality and women's self-identity. In Cloke P. and Little J. (eds), *Contested Countryside Cultures: Otherness, Marginalization and Rurality*. London, Routledge, pp. 138–157.

Little, J. (1997b), Constructions of rural women's voluntary work. *Gender, Place and Culture* 4 (2), pp. 197–209.

Little, J. (2001), *Gender and Rural Geography*. London, Addison Wesley Longman.

Little, J. and Jones, O. (2000), Masculinity, gender and rural policy. *Rural Sociology* 65 (4).

Myers, F. (1999), *Women in Decision-Making in Scotland: A Review of Research*. Women's Issues Research Findings No. 2, Edinburgh, Scottish Office.

Jones, N. (1997), Diverging voices in a rural Welsh community. In Milbourne, P. (ed.), *Revealing Rural 'Others': Representation, Power and Identity in the British Countryside*. London, Pinter, pp. 135–145.

Jones, O. (1999), Tomboy tales: the rural, nature and the gender of childhood. *Gender, Place and Culture* 6 (2), pp.117–136.

Seymour, S. and Short, C. (1994), Gender, church and people in rural areas. *Area* 26, pp. 45–56.

Shortall, S. (1994), Farm women's groups: feminist or farming or community groups or new social movements? *Sociology* 28 (1), pp. 279–291.

Shortall, S. (1996), Training to be farmers or wives? Agricultural training for women in Northern Ireland. *Sociologia Ruralis* 36 (3), pp. 269–285.

Shortall, S. (1999), Political restructuring in Northern Ireland: Implications for equality proofing CAP reform measures. Paper presented to the *Gender and Rural Transformations in Europe* conference, Wageningen, Netherlands, October 1999.

Stone, M. (1990), *Report on Rural Childcare*. Research Report No. 9. Rural London, Development Commission.

Symes, D. (1991), Changing gender roles in productivist and post-productivist capitalist agriculture. *Journal of Rural Studies* 7, pp. 85–90.

Whatmore, S. (1991), *Farming Women: gender, work and family enterprise*. London, Macmillan.

Whatmore, S. (1993), Agricultural geography. *Progress in Human Geography* 17 (1), pp. 84–91.

Whatmore, S., Marsden, T. and Lowe, P. (1994), Introduction: Feminist perspectives in Rural Studies, In Whatmore, S., Marsden, T. and Lowe, P. (eds), *Gender and Rurality*. London, David Fulton, pp. 1–10.

Woodward, R. (1998), 'It's a Man's Life!': soldiers, masculinity and the countryside. *Gender, Place and Culture* 5 (3), pp. 277–300.

Woodward, R. (2000), Warrior heroes and little green men: military training and the construction of rural masculinities. *Rural Sociology* 65 (4)

Chapter 10

Rural Gender Studies in The Netherlands

Bettina Bock and Henk de Haan

This chapter describes the development of gender research on agricultural and rural development in the Netherlands since the 1950s. The aim of the chapter is to analyse how the concept of gender appears in different periods of rural research. Academic discourse on rural gender issues cannot be analysed without taking into account its social, cultural, academic and political contexts. In the Netherlands academic research on gender issues in agriculture and rural society has mainly developed within the field of rural social studies, rural sociology in particular. In order to understand the way in which gender has been approached, it is important to know that Dutch rural sociologists have always cooperated intensively with rural interest groups and policy-makers and in doing so have attached great importance to the practical applicability of (academic) research. As a result, academic and political debates are 'intertwined' (Nooij, 1997), and research agendas and theoretical orientations in rural sociology are developed in interaction with policy needs and the social and political definition of problems. The development of rural gender studies reflects this discourse coalition. When considering gender research on agricultural and rural development, it appears that the academic and social-political domain are so much connected and scientific and policy discourses have so many common characteristics that it is not easy to distinguish between the definition of social and academic problems. Therefore it seems appropriate to present rural and broader socio-political developments simultaneously with the emergence of academic research agendas, and to consider the development of rural gender studies within this context.

This chapter presents a chronological overview of rural gender studies. The periodization is based on parallel developments in rural studies, gender research, agricultural and rural developments, and social, cultural change in society at large. From this perspective we distinguish three phases. *Phase one*, which coincides with the classical modernization period, covers the time between around 1950 and the end of the 1970s. It is characterized by a denial of farm women's productive role, and a concept of emancipation that identifies female qualities with the household, the family and civil society. The *second phase*, which lasted until the mid-1990s, marks a period when farm women were rediscovered, redefined and general emancipation ideals were translated into the agrarian context. This period coincides with the emergence of critical social movements and the beginning of a general crisis of the modernization model. The *third phase*, which started in the mid-1990s and continues today, is more difficult to characterize. Much of what was going on in the earlier periods has continued, but gender studies has also broadened its scope. While it used to focus on gender and agriculture, and mostly on the distribution of labour and decision making within the farm and farmers' organizations, attention is now shifting towards rural women's participation in the broader social-political process of rural development and countryside renewal. While in former periods the effects of broader changes on the position of men and women were emphasized, now the active contribution of women to these changes has become more the centre of attention.

Farm Women Marginalized: The Invisible Agricultural Labour Force (1950–1980)

In the 1950s farmers still dominated village life and most of the land was used for agricultural purposes. Industrial activities and services were concentrated in urban settings. Farms were generally small or medium-sized, although the farming structure differed considerably by region. Especially in the sandy-soiled regions in the South and the East of the country, large numbers of small farms could be found. Government and agricultural organizations both considered a reduction of the agricultural labour force and the number of farms necessary, in order to raise labour productivity and to enlarge the scale of viable farms. At the national level the government aimed at eliminating economic and socio-cultural differences by integrating all regions into a welfare state with equal opportunities and rights. In general, agricultural and regional modernization were not defined in terms of overcoming technical and economic obstacles, but rather as a battle against traditionalism. The family farm was considered the main fortress of traditionalism, not only in an economic sense, but also in social and cultural terms. The seemingly 'irrational' mix of family, household and production, and the associated sense of belonging to the land and the village were the main targets in government policy aiming at transforming agriculture into a modern economic sector (De Haan, 1993).

This modernization policy was not restricted to farming alone. It was strongly supported by the pursuit to build up a welfare state, which was not only defined in terms of social security and employment, but also in cultural terms. It involved fundamental ideas about the role of the family as the cornerstone of society. Civil society was considered to depend upon the formation and education of responsible people, with a stable and healthy family life. From this perspective women performed a crucial role, both as mothers and as wives. Married women's participation in the labour market was condemned as being harmful to family life and detrimental to the health of children.

This civilizing and conservative bourgeois ideology also had its impact on the countryside, where the cult of domesticity was introduced in special educational programmes, extension services, health care and other social provisions (Verboon, 1988). Together with the government's rural modernization policy, this had important consequences for farm women in particular. As later (historical) research revealed, it led not only to a lightening of the often heavy farm work women were used to doing, but also to a serious degradation and deskilling of their professional role (Van der Burg and Lievaart, 1998). The liberation of women from heavy and dirty farm work was seen as a condition for their engagement in civilizing rural people and rural society. Women needed more time for the education of their children, to reorganize the household to meet bourgeois standards, and to engage in voluntary projects in order to improve the development of rural areas (Van der Burg and Lievaart, 1998; Verboon, 1988). Farmers' wives were in an ambivalent position in this changing ideological context. How could they meet the standards of good housekeeping and motherhood if they had to work fully or part-time on the farm? And how could farmers dream of separating farm work from family work if they relied heavily on unpaid family labour?

The Marginalization of Women from Farm Work

Academic research during the 1950s and 1960s largely ignored this female factor. Policy-makers simply were of the opinion that the modernization process takes time and if reality did not exactly correspond with ideology, this was explained by the force of tradition and cultural retardation. Studies on family farming by Kriellaars (1951) and Saal (1958), which both emphasized the crucial labour role of women, were considered as historic documents, not relevant for the visionary future of a male-dominated entrepreneurial sector. During the period from 1950 to the mid-1970s academic discourse on the position of women in farming and rural life was almost non-existent. Women were invisible actors, and as far as their existence was acknowledged, it was not from the perspective of farm work or other productive roles, but as mothers and wives (De Groot, 1989; Van der Burg and Lievaart, 1998; Van der Burg, 2002).

On very few occasions farm women appeared in academic or policy-oriented studies. Kriellaars (1951), for instance, a conservative observer, praised the family farm for its moral superiority and its capacity to unify the whole family in production and consumption. He saw the patriarchal structure of family life as a guarantee for stability, and the submission of women to men logically implied their availability for farm production. The family sociologist Saal (1958) conducted research on farm families, in which he developed a classification of family farms on the basis of women's role in the labour process. He distinguished several organizational patterns, ranging from complete separation to complete incorporation. His analysis shows the close connection between farm and family and between productive and household labour. Nonetheless he observed a trend towards the marginalization of women from the farm and their confinement within the household, which was vividly illustrated by the transformation of cheese making from a home to a manufacturing industry. His concern with the changing nature of the family reflects a wider tendency within Dutch sociology of characterizing family development in terms of a process of functional specialization, resulting in sharply distinguished spheres of production and consumption. This specialization process did not take place on all farms, however. In some regions, especially in the West of the country, farm women managed to continue the on-farm production of cheese, even until today. Kooy (1956) explained this persistence as an attachment to regional culture and tradition. In later historical research it was described, however, as active resistance of women to the forces of modernization because of their love for and pride in their cheese making profession (Bock and Van der Burg, 1998; Van der Burg and Lievaart, 1998).

Retrospectively, cheese making did indeed gradually disappear from most (though not all) farms. Interestingly enough, no further research was done at that time to document and analyse the new labour relations coming in its place on modern dairy farms. This lack of interest underlines again the dominant modernization perspective from which these studies on women's farm roles were performed. From that perspective further research was indeed not necessary: a general rise in affluence and modernity in the agrarian sector would unavoidably lead to functional specialization, with women disappearing into the private domain of the household and the family.

In the same period new statistical methods were developed for measuring farm labour: in the 1950s farms below 1 ha were no longer included in official statistics and in the 1960s irregular work and a labour input of less than 15 hours per week were no longer registered. Both resulted in female farm labour becoming invisible and confirmed the general idea that women were no longer involved in serious farm work, whereas in reality only few (bigger) farms were able to cope without collaborating spouses (Bock and Van der Burg, 1998).

The Feminization of Farm Women

While women were politically and ideologically invisible as labourers on the family farm, their roles in family life, and rural society were clearly emphasized (see Van der Burg and Lievaart, 1998). A good example of this is Stork-van der Kuyl's (1952) study on farm women. Although she described the tasks of farmers' wives in detail, she was mainly interested in their 'emancipation' from the isolated worlds of farm and community. Farmers' wives, she argued, ought to be part of such emergent cultural patterns as the process of 'feminization,' i.e. '. . . the development of the typical female element; the charming features, which could not develop in former daily life' (p. 192). She described how women's attention was increasingly focused on family and household, on problems of housing, clothing, food, hygiene, and education. Emancipation for this author referred to women freeing themselves from the hard work on the farm and devoting themselves to more 'feminine' and more rewarding tasks. She therefore proposed that farm women specialize in bookkeeping, which would also grant them more insight in the management of the farm as a whole. The liberation from hard farm labour would enable farm women to develop themselves as active agents in the 'civilizing' process.

It is true that in Dutch society in general this version of female liberation was quite successful. In the 1950s and 1960s female participation in the labour market was minimal. In 1960, 6.8 percent of married women had paid employment and until 1955 women working in public services were forced to resign on marriage. At the same time an impressive infrastructure was developed to help and educate women in becoming perfect 'modern' housewives. Specific courses and extension programmes were developed for rural women in order to disseminate information about nutrition, budgeting and furnishing: in short, the equipment and organization of modern rural households (Verboon, 1988; De Groot, 1989). The general ideological climate thus made women invisible farmers. Academic studies on agriculture and rural life generally ignored women, focusing on structural and social problems related with farm modernization. At most, some evaluating studies were done in order to examine the adoption of modern household methods by rural women (see De Groot, 1989).

Only in the early 1970s was some attention given to farm women's role in the modernization process. It was researched how women influenced decision-making in farming, and which attitudes they had about leaving farming. These studies (Spierings, 1972, 1974; Kloprogge, 1973) showed that women's attitudes may have been very important, especially with respect to their desire to increase farm incomes in order to satisfy new consumption and lifestyle demands. But the research also showed that women felt attached to the agricultural profession and would experience leaving farming as a loss. In general, however, men took the final decision about occupational change.

The Discovery and Empowerment of Farm Women in the 1980s and 1990s

In the late 1970s the modernization project had been 'successfully' completed and institutionalized, at least from a policy point of view. Thousands of farmers had left agriculture, and farms were firmly integrated into the agribusiness complex. The close cooperation of education, extension and research resulted in an enormous increase in productivity and the specialization of Dutch agriculture in so-called bulk production: farm products of reasonable quality against low prices. Farming was more and more considered a 'regular' profession and a business, requiring specialized vocational education and the acquirement of entrepreneurial skills. With the closing down of many small farms the average size of farms had considerably increased. Nevertheless, family farming had not disappeared, but proved to be an organizational structure well equipped for incorporating entrepreneurial practices.

The completion of modernization confronted academic rural sociology with an identity crisis (De Haan and Nooij, 1985). Its 'golden age,' which rested upon a close connection between policy and theory, came to an end, as it was firmly embedded in the socio-political paradigm of modernization that began to lose its significance and lost most of its appeal among a younger generation of sociologists. The expansion of agribusiness and the process of vertical integration shifted attention towards chain management and the dependent role of primary producers. Moreover, overproduction, environmental problems and the devastating results of agricultural restructuring on landscapes increasingly drew attention to the drawbacks of what was called the 'agricultural treadmill.' In this context, academic research was beginning to define a new research agenda, and in this process it initially lost its close connection with government policy-making.

Women's Movement and the Rediscovery of Farm Women

This 'identity crisis' in rural sociology, together with the appearance of new social movements occupying themselves with emerging agricultural and rural problems, paved the way for new perspectives. Apart from new interest groups, such as the very active young farmers' organization, and other critical agricultural movements, an important factor in the redefinition of the research agenda was the Wageningen student movement. Within that movement a group of activists was very concerned with the marginalization of farmers in agribusiness, the one-sided support for scale enlargement and the disappearance of small farmers. The movement was especially interested in the personal fate of farmers in the face of capitalist expansion in agriculture. Part of this movement focused on farm women. This seemed a logical translation of what was going on in society at large. The 1970s witnessed the birth of the second feminist wave and the growth of a strong women's movement arguing for equal opportunities in work and politics and the right of self-determination. The changing position of women in society was reflected in the growing number of married women in the labour market, which increased to 26 percent by 1979.

These general feminist concerns were projected on rural society and this resulted in the rediscovery of farm women. Farm women had a lot in common with housewives, who formed the massive group of the unpaid domestic and likewise 'invisible' labour force. The situation of farmers' wives was however much more complex. They were in fact taking part in unpaid productive work, and as such their position was in between women with paid employment and those without. Demands for equal participation in the labour market and equal wages were thus not quite applicable to the farm context. But before any such theoretical or political questions were asked, the first step towards formulating women's demands was to discover this invisible labour force: who are these farm women, what is their position, to what extent do they participate in farm labour? Initially this politically motivated interest resulted in a number of student research initiatives, but the Ministry of Agriculture's research institute quickly picked up the ideas.

Female students of the Agricultural University, often farming daughters themselves, started to look more closely at the amount of farm work women actually performed (Hobbelink and Spijkers, 1986a,b). Most of them were social scientists, specialized in rural sociology, rural household studies or (agricultural) economy. But (female) students from other universities also joined in, most of them anthropologists and historians. They revealed that, despite general expectations, women were increasingly collaborating in farm work, especially on dairy farms (see Bos, 1969; Bosch, 1970; Litjens et al., 1979) because of production expansion and the growing administrative and general management workload (Hoerchner et al., 1979). It thus seemed that farm modernization had not produced a complete separation between private and productive spheres.

Farm Work and Decision Making: Women are Made Visible

The first national and representative research on the condition of farm women was published in 1984 (Loeffen, 1984, and repeated in 1989 by Blom and Hillebrand, 1992). Its results gave a detailed, quantitative picture of women's participation in the labour process and their role in farm decision-making. Most farm women appeared to have a double workload, spending on average about 20 hours per week doing farm work, besides being fully responsible for the household. The discovery that women were responsible for such a significant input in the total agricultural labour force did not come as a surprise to farm women and farm interest groups, but research elevated this fact from personal experience and subjective knowledge to an economic and political fact. Although this knowledge was not immediately translated into policy recommendations and theoretical reflections on gender, the simple inventory of female participation was an important step towards putting farm women on the academic and political agenda. It bolstered female consciousness in the sense that women no longer reflected on their situation in private terms, but as a shared experience. It also proved the inadequacy of official labour statistics, which overlooked and ignored a significant part of the work actually done on farms (Van Wijk, 1983; Bock and De Rooij, 2000). In the same period a lot of historical research was done to discover the significance of women's role in agriculture in the past and to analyse and explain the change that occurred over time (Backerra et al., 1989; Van der Burg, 1988, 2002; Van der Burg and Lievaart, 1998).

Even more important than the purely quantitative inventory of female labour participation were the outcomes concerning the kind of tasks women performed and their participation in decision-making. Loeffen (1984) clearly demonstrated that women were hardly involved in specialized skilled work, but rather performed a varied collection of small tasks. This 'helping out' was typical for the flexibility of farm women: being available meant that their assistance could be mobilized at all times. Loeffen also differentiated between work *on* the farm and work *for* the farm; the latter included all kinds of errands like the laundry, making arrangements with the veterinarian and bookkeeping. Especially the work *for* the farm was rarely recognized and accounted for as 'real' work. Concerning decision-making, the research showed that women were very much concerned with the farm. Spouses discussed and deliberated plans about investments and other long-term developments, but men generally made the final decisions. In other words, research underlined the importance of women's input in the farm but also revealed the weakness of women's position and status. In general women had a supporting role, without full responsibility for the labour process or farm management. Farmers' wives were not considered to be professional farmers or (co-)entrepreneurs, but first of all collaborating spouses, whose farm work was taken for granted as the consequence of being married to a farmer and running a rural household (Aarnink, 1987; Alkemade and Alting, 1995). To fulfill this natural obligation without complaining and without asking anything in return was also considered to be the appropriate behaviour and attitude of a good farm woman (De Groot, 1989).

Research for Farm Women: Policy Orientations and Social Problems

As mentioned above, the theoretical aspirations of this first research wave on farm women were limited. On the whole, research was based on the idea that there is a certain amount of farm work, which can be divided into different tasks and skills and finally results in different roles. Important insights were derived from the recognition that current definitions of work and the use of these definitions for classifying productive and non-productive tasks have a normative character. By redefining the concept of labour in the context of family farms, women's fundamental role could be made visible. But initially, little attention was given to how these roles are structured and reproduced in daily interaction and to what extent they are embedded in gender ideologies and are typical for family-based agricultural production. The fact that such theoretical questions were hardly developed and that research in later years mainly focused on policy questions in cooperation with women's organizations is no surprise. While rural sociology in general was still trying to find a new position *vis-à-vis* the turbulent developments in agriculture, gender studies soon institutionalized as a client-oriented academic practice. Thus, research in the 1980s and part of the 1990s reflected the policy agenda and interests of farm women and farm women's organizations, rather than the international academic agenda concerned with problematizing gender in a theoretical way (Hobbelink and Spijkers, 1986; Jansen, 1991). As a result, research results were mostly disseminated by way of research reports and popular magazines.

In the 1980s farm women's emancipation was defined as something to be achieved both at the farm level, and in the public sphere. At the farm level the issues were related to achieving more respect and equality, or empowerment, mainly in the division of farm labour and management. In the public sphere women's interests focused on improving their legal and fiscal position, and on a better representation of women in farm organizations. In general, women's position was defined in terms of subjection, low status, marginality and exclusion.

In this period a lot of (policy-oriented) research was done for various farm women's organizations and groups. It concerned the development of methods for making women's work visible and thus accounted for (Dorresteijn, 1983; Lodder and Tims,

1988; Berger et al., 1987; Van Wijk, 1986; Blom and Van de Hoek, 1988), the effective organization of farm women in separate pressure and lobby groups, and the integration of women into mainstream farmers' unions (see Haenen, 1992; Jansen et al., 1985; Pijneburg, 1991; Brand and Verhagen, 1991; Bosch, 1990; Van Poppel et al., 1990; Van Wijk, 1986; Litjens and Klaver, 1991). It is interesting to note that farm women could personally become union members only from 1984 onwards. Before that time union membership was limited to farm principals, who were generally men. Moreover, research was done on women's chances of entering governmental policy-making networks, such as land consolidation committees (Blom et al., 1986; Van den Hoek et al., 1987; Keuzenkamp, 1997). Other researchers focused on social security issues and health (Bloemhoff and Wolf, 1987; Bun et al., 1990; Bun, 1991; Blom and Van de Hoek, 1989; Klein et al., 1986; Melita, 1993; Jansen, 1990; Lieshout, 1989; NAJK, 2000), also in relation to the deteriorating situation in the agricultural sector (Giesen, 1991; Van Zwieten, 1994), and the improvement of the legal, fiscal and social position of farm women (Lodder, 1991; Overbeek, 1993). The latter resulted in the proposition of an official professional status of collaborating spouses, which was never approved, however (Lodder, 1991). Instead the government adopted and fiscally promoted the option of business partnerships between spouses (Overbeek, 1993). Furthermore, general research was done on education and extension for farm women (Van der Burg, 1988; Van Doorne-Huiskes and Rosendahl, 1998), in particular on the position of girls in agricultural education (Sinnema, 1990; Laval and De Rijk, 1990; Kools, 1994, 1996; Van der Wal, 1990). As can be noticed from the choice of topics, this type of policy-oriented research is still relevant today. This is especially true regarding the integration of farm women into farm unions and agricultural policy formation networks (Van Casteren, 1997; Heel and Lodder, 1994; Hilhorst, 1993).

Theoretical Approaches: Power and Identity

Most of the research on gender and agriculture focused on policy issues and may be characterized as applied research. Nevertheless some important theoretical advances were made in that period too. As most of the theoretical achievements were the result of dissertation research, we take these as a guideline. We present them in sequence of publication, while also giving references to other (prior and later) research offering theoretical elaboration on the same or related subjects.

In 1990 Zwart, specialized in rural household studies, published her dissertation on women's position within the family farm. She developed her theoretical framework within the context of family sociology, focusing in particular on the effects of farm modernization and the individualization of family relations on the position of women. Her research revealed that labour and family, working and living, the public and the private were still closely interwoven in family farming, although in society at large public and private spheres were increasingly separated (see also Hobbelink, 1982, 1991; Klaver and Zwart, 1996). She particularly emphasized the intermediate

role women fulfilled between family and farm and the function of women as surplus labour. If necessary, women were expected to help on the farm, which they had
to combine in a flexible way with their responsibilities for the household and family.
Women's involvement in farm work was heavily influenced by the necessities of the
farm and her responsibility for the household and children (which depended on the
family cycle). If for instance the successor was entering the farm, she was often replaced, with respect to farm work and farm management, by her son, who became the
new partner of her husband, generally without even discussing other options (Jansen, 1990; Klaver and Van Poppel, 1994). The taken-for-granted interrelation of family
and farm also resulted in the seemingly natural priority of business interests, because
'what is good for the farm is good for the family.' This also explains the ambivalent
opinion of many farmers (and farm women) concerning the representation of farm
women's interests in farm unions (see Van Wijk, 1996). Farm women and men were
considered to share the same interest: the prosperity of the farm.

Giesen (1991) examined the integration of farm and family from a psychological point of view and showed that conflict and considerable stress could occur during periods of economic and private problems. Other theoretical contributions were
made by the anthropologists Sweere (1991) and Harpe and Potters (1991) who, focusing on the position of (future) female farm operators, drew attention to how women
shape their professional identity and create space by mobilizing specific sources of
power. Women are not merely passive victims, but are also actively defining and redefining female roles. According to Sweere (1991), for instance, women consciously
try to construct and hold on to 'femininity' in several ways, while challenging at the
same time male roles and identities. Their room for maneuvering is, however, limited
by gender-specific norms and values.

In her dissertation on women on dairy farms the rural sociologist De Rooij (1992)
explained the subordinate position of women within the farm as the outcome of the
modernization process. She argued that the position of women within the farm and
in farm management depends on how the labour process was organized and adapted
in the course of modernization (De Rooij, 1992, 1994). Her approach was much inspired by theories about the quality of labour and labour task control (Mok, 1990). In
her view modernization resulted in the gradual elimination of female labour domains
because of the externalization and mechanization of typical female tasks. As a consequence, women's labour was de-skilled and degraded. What finally remained were
mainly dissociated and dispersed 'helping-out' tasks, meant to assist her husband under his supervision and guidance. This again confirmed the hierarchical character of
their working relations and women's decreasing decision-making power. De Rooij's
research convincingly demonstrates the gender-loadedness of this process, as the
degradation of part of the farm work took place only when the cooperation of men
and women was concerned. Comparing the labour relations between spouses on the
one hand, and male business partners (fathers and sons) on the other, she revealed
that male partners were always careful to reach an equivalent and equally rewarding

division of farm labour and management. When the division of labour between men and women was concerned, arranging a balanced situation was obviously thought to be unnecessary. Especially on entering the farm as newlywed wives, women mostly took on those tasks not yet claimed by the husband or parents-in-law and which nobody else was willing to do (Klaver and Van Poppel, 1994).

Gender Studies and Rural Studies

In summary, the 1980s and 1990s saw a growing interest in farm women. First women were made visible, then their private and public positions were challenged, and finally theoretical work on power relations and the construction of female and male roles placed this empirical and policy-oriented work in a broader perspective. In general, research moved slowly from an empirical and policy-oriented focus on women's secondary roles and asymmetrical power relations, to more theoretically oriented research. In doing so it created a perspective on how women's roles are connected with farm and family developments and on how women create space in trying to find their own identity and challenge traditional roles. In some respects research on women was unique in the context of rural social studies, as it focused on the human factor. As such, gender studies more or less paved the way for a more actor-oriented approach in rural studies. By connecting individual experiences, attitudes and changing social relationships with such abstract notions as structural and political change, the human dimension was fully exposed. In general rural sociology this perspective also proved successful, for instance in the farming styles approach by Van der Ploeg (1999), although this approach mainly focuses on the farm entrepreneur (for an exception see De Rooij et al., 1996). As all the researchers engaged in rural gender studies are women and many of them have themselves experienced a male bias in science, they are, perhaps, more inclined than men to recognize the professional status of farm women as well. From a methodological point of view, most of their research may be categorized as qualitative or mixed, using mainly un- or semi-structured personal interviews and some quantitative data and statistical analysis.

Rural Women and Rural Development Participation in the 1990s

During the 1990s the situation in agriculture became more and more difficult in economic as well as political terms. The globalization of production and trade and increased competition resulted in a lowering of prices. At the same time production costs were increasing because of stricter regulations concerning environmental protection, food safety and animal welfare. The national and European subsidy schemes offered less production-related remuneration than they did before. In general and very briefly summarized, European and national policy-makers tried to shift the focus of agricultural policy from primary production and maximal productivity towards rural development (or rural renewal) and multifunctional and sustainable farming. At the same time a growing number of farm households began looking for new and additional sources of income. They initiated new activities on the farm, such as the transformation and direct sale of farm products, nature care, or offering different kinds of services, mostly in tourism and recreation, but also in education and health and child care. In rural studies this is defined as part of the transition of rural areas from production to consumption and the adoption of a new, endogenous paradigm of rural development (Van der Ploeg et al., 2000). Furthermore, farmers tried to adapt production orientations and methods to the new social and consumers' demands. As a result organic and high-quality production was growing and farmers increasingly participated in different kinds of chain management and control systems. In this way they tried to regain consumers' confidence, which had been very seriously damaged by various scandals (BSE, FMD, dioxin, nytrophene). Moreover, a growing number of farming men and women began combining farming with a job off the farm in order to be able to cover household expenses, to finance further farm investments and to secure the continuity of the farm. The decrease in the number of farms and the increase in average farm size continued, as well as the tendency towards further industrialization. But modernization was no longer the only and generally promoted or accepted path of development. For a growing number of medium-sized farms in particular, the turn towards quality production, farm diversification and pluri-activity offered a new and promising future (Van der Ploeg et al. 2002).

During the 1990s the government intensively promoted economic independence for women by stimulating girls to invest in a solid vocational education (preferably in science) and by convincing young mothers to either not leave or return to the labour market soon after childbirth. Apart from large-scale promotion campaigns, the adoption of fiscal laws was used as an important policy instrument. The participation of women in the labour market increased constantly and remarkably: from 44 percent in 1994 to 62 percent in 1999 (Bock, 2001). In rural areas the same development can be seen, although the participation rate of women remains lower compared to urban areas. In the agricultural sector a persistent growth in the number of female (co-)farm heads can be witnessed, from 13 percent of all farms in 1993 to 26 percent in 1999 (Overbeek, 1993; Bock and De Rooij, 2000).

Broadening the Research Agenda and Farm Women's Horizon

Most of the rural gender studies research themes from previous periods were continued during the 1990s. But there were also some new developments. First of all, the issue of property and its unequal, gender-specific distribution received attention. Whereas in so-called developing countries the exclusion of women from farm property has long been an issue of research and policy-making, the situation in Dutch agriculture has not received any attention so far. Most likely this can be explained by the persistent farm family ideology. Recent research reveals the unequal distribution of property between men and women (Zwart, 1996; Overbeek, 2000) and the mechanisms by which this inequality is constructed and even increased by the recent practice of marriage contracts (Bernet, 1998). What was initially meant as a construction for protecting the farm from risks (like divorce), actually maneuvered farm women into a more insecure and exploitative economic relation than in former times, when couples used to be married in community of property and benefits (Zwart, 1996).

Furthermore, research moved from farm women and their role in farm work, to women's participation in society at large and their role in off-farm labour markets. Although farm women's rural residence and background are still important for the construction of gender roles in a rural setting, this trend constitutes a break from defining farm women only on the basis of their status as farm wives. Especially in the 1990s it became clear that young women no longer took it for granted that they should be available for farm work after marriage. It is in fact a paradoxical development that so much research energy was invested in improving women's status on the farm and their role in professional organizations, while at the same time young women turned their back to the farm in growing numbers. Although they remained of course closely involved in farm life, they increasingly seemed to define their professional role not within, but outside agriculture, and not in partnership with their husband, but independent of him. In 1992, Blom and Hildebrand had already demonstrated the changing attitude of young farm women, who increasingly participated in off-farm work and thus followed an individual and independent professional career outside the farm (see also Overbeek, 1995; Overbeek et al., 1998).

The economic insecurity of farm families is one factor that explains why women enter the labour market; the high costs connected with the transfer of the farm is another. Moreover, the growing popularity of father-son partnerships as a mode of transferring the farm gradually is a factor that is pushing women out of farm labour (Hildebrand and Blom, 1993; Klaver and Van Poppel, 1994; Jansen, 1990). Besides, a growing number of farm women do not have an agricultural background, and as they are not really involved in their husband's business and profession they orient themselves away from the farm and do not consider themselves 'real' farm women (Bock and De Rooij, 2000).

However, the research agenda is expanding in a different way as well. This is closely connected with the developments in rural areas described above and the increasing importance of diversification and pluri-activity. At the end of the 1990s research on the role of women in rural development resulted in several studies (Hendrikse and Klaver, 1995a,b; Bock, 1996, 1997; De Rooij et al., 1995; Bock and De Rooij, 2000; Overbeek 2001). These studies revealed that women are keen on starting new activities on the farm, not only for generating extra income but also as a strategy to create their own labour domain on the farm and an independent professional identity (Bock, 1996, 1997). It is also a way of (directly or indirectly) gaining more influence on the management of the farm as a whole (Bock, 1994, 1998). Furthermore, women are generally more convinced than their husbands that diversification offers a promising future for their farm (Bock and De Rooij, 2000). They more often reject the idea that further expansion, intensification and industrialization of agriculture are viable options for the future of the Dutch agricultural sector as a whole (De Rooij et al., 1995).

Research on gender-specific differences in innovative behaviour demonstrates that women tend to start cautiously on a small scale in order to limit the costs of investments. This allows them the option of giving up if the business fails, or if the combined workload of farm, family care and the new business is too intense for them or their children (Bock, 1998, 1999). Although women tend to accelerate professionalization after this cautious beginning, they have great problems in getting any governmental subsidies. Generally they can only benefit from set-up subsidies, which are only given on the condition that there is a relatively expensive, one-step launch of a business. It is clear that such subsidies are not adapted to female enterprise development strategies and are basically designed for stimulating 'male' innovative behaviour (Bock et al., 2000).

Rurality, Gender and Power

Apart from women's role in new economic activities on and off the farm, other aspects of women's participation in rural development have also been studied. Among them are the involvement of women in rural development projects, their access to rural development subsidies and programmes like LEADER, and their involvement in the process of interactive policy-making (Bock, 1998, 1999, 2002). These studies underline the gendered nature of the process of rural development and the partiality of women's involvement. In practice, women are important agents in the social and economic restructuring of the countryside; they are, however, nearly completely neglected when the development of plans and policies is concerned. The model of endogenous rural development, which guides scientists as well as policy-makers, thus results in practice in a selective participation and empowerment of men and women. At a more abstract level, this points to the gender-blindness of the actor-oriented theories of endogenous development as well as (participatory) governance and policy-making (Bock, 2002).

So far, little attention has been paid to the gender-specific assessment of the quality of rural areas and the perception of rurality. The only exceptions are some policy-directed studies on the quality of life in rural areas and the specific opportunities and constraints they offer to women, especially concerning their participation in the labour market and the combination of work and child care (Janssen and Lammerts, 1999). These studies in fact continue a long-standing tradition of research on rural instead of farm women (including non-farming women). It includes issues like participation in organizational boards and voluntary work, access to (local) labour markets and constraints on emancipation in the specific social (Brunt, 1975) but mainly geographical characteristics of rural areas (see Oomens and Kriekaard, 1989; Lohuis et al., 1987; Ten Brinke-van Hengstum and Tonen, 1988; Drooglever Fortuijn et al., 1994; Bettenhausen-Verbey and Blom, 1983; NIROV, 1990).

So far, no attempts have been made to explore and explain the gendered contents of rurality as a socially constructed concept and discourse. In part this probably can be explained by the general lack of interest of scientists as well as policy-makers in the issue of rurality and the specific socio-cultural identity of rural areas. Until very recently it was generally assumed that with modernity the differences between rural and urban areas and city and country people would vanish. And even geographical constraints are considered to lose importance in view of the generally increasing mobility of people and the opportunities ICT is offering. As a matter of fact, official statistics no longer differentiate between urban and rural areas, which makes it impossible to compare the labour market participation of urban and rural women. Moreover, it is interesting to note that while formerly the research on farm and rural women used to be rooted in different disciplines and universities, a certain rapprochement can be noticed today (see Bock, 1998). The rising consciousness of the shared ownership of the countryside by all inhabitants (and not only farmers) and thus the common interests of all rural women concerning the future of their living environment and way of life is certainly playing a part in this. Maybe this development will encourage researchers to pay more attention to the subject of rurality and its gendered nature.

Conclusion

The development of rural gender studies in the Netherlands during the second half of the 20th century shows that most attention was focused on women's role in farm households and that this offered a strong ideological and political dimension in emphasizing the need for change. In the 1950s women's combined role as housewife, labourer on the farm and spouse was considered a heavy burden and detrimental to the welfare of the household and the potential role of women in local society. Emancipation was defined as liberation from farm work so that women could devote all their time to the family and the community. The separation of production and consumption was inspired by the recognition that women's specific qualities lie in caring.

In the early 1980s it became clear that this emancipation ideal had not resulted in a masculinization of farming. The agricultural modernization process strongly depended on women's farm labour input. In the context of wider feminist movements this observation called for a recognition of women's productive roles and a redefinition of male–female relationships, both in the private sphere of the household and the public sphere of interest representation. From emphasizing a woman's roles as housewife, mother and spouse and stressing her female qualities, her role in the labour process became central. Emancipation ideals were expressed in terms of empowerment, equal representation and equal rights. Much research was devoted to documenting women's position on the farm and how it could be improved by new policy measures. By emphasizing the gendered nature of power relations, women were frequently portrayed as the victims of a male-dominated agricultural world, struggling to emulate role models that were based on the model of masculinity. This may well have contributed to the fact that much less attention was given to emergent heterogeneous patterns of ways in which men and women construct ideas about femininity and masculinity. The dominant view on gender relations did not leave much space for exploring differences between women and the space women created for themselves for defining their own identity.

The last phase in rural gender studies described in this chapter draws attention to the broadening of focus from the farm to the wider rural environment and to women's active role in rural innovation. While in the previous period women were mainly viewed as objects, or persons aversely affected by rural change and gender ideologies, they are now increasingly viewed as subjects: as persons from which the action proceeds. This is in particular emphasized by research on the role women play in rural entrepreneurship, but it is not yet clearly demonstrated in the public policy sphere. It seems that the recent government shift towards regional and local interactive policy processes has revealed a new field in which women seem to take a disadvantaged position *vis-à-vis* men.

Looking back over the whole period of rural gender studies, it seems clear that the strong emphasis on farm women and later on women as entrepreneurs has left a vacuum in our knowledge about the relationship between gender and rurality. Now that we have arrived at a point where it is recognized that both rurality and gender are social constructions, it seems time for more grounded theoretical and empirical work on how gendered rural identities are shaped by concrete practices and discourses. This is particularly relevant for future and present policy concerned with rural areas. Recently the Dutch ministry responsible for rural areas announced that rural development should be built and based upon 'social cohesion'. Social cohesion, or social capital, should provide the basis for local initiatives, political participation and responsibilities for environmental quality. It is very important that this sort of rhetoric not results in a marginalization of those who have traditionally been excluded from policy-making.

A final observation concerns the relevance of this overview in an international perspective. In many respects the developments in the Netherlands are not unique. Most European countries have experienced the described phases of modernization and professionalization of agriculture, followed by a trend towards more diversification in recent decades. It seems that during the classic modernization period the role of farm women was generally denied, and that gender studies contributed substantially to the discovery and re-valuation of productive work in a domestic context. With the professionalization of agriculture women everywhere demanded an equal status, and policy claims are framed in terms of empowerment. With the emergence of the 'consumption countryside', including the enlargement of rural labour markets, women are taking the opportunity to reclaim and claim space for defining their own identity, thus contributing to the diversification of rural space. It is difficult to assess what is typically Dutch. Perhaps one feature is the relatively late appearance of farm women on the labour market. This is not only connected with a long-standing negative attitude towards female labour market participation in general and in rural areas in particular, but also with the fact that agricultural policy strongly discouraged pluri-active farm households. A more prominent role on or off the farm did not match the model of the professional farm enterprise. Another point that may be typical for the Netherlands is the general lack of a female, or feminist, perspective on issues of broader rural relevance, such as nature, tourism, planning and policy. It seems to be very difficult in Dutch rural studies to develop a research agenda and an academic discourse, independent from the policy agenda. Dutch rural studies seems to be most adept at playing a role in a discourse coalition, where policy-makers, academics, civil society and ordinary people debate the usefulness of policy alternatives, leaving less space for intellectually challenging projects.

References

Alkemade, S. and A. Alting (1995), *Beid'r belang. Een onderzoek naar de positie van de vrouw op het agrarische bedrijf*. Tilburg, NCB.

Aarnink, N. (1987), *Mijn man werkt en ik help mee. Veranderende familie-, arbeids- en man-vrouw verhoudingen binnen agrarische gezinsbedrijven in Nederland*. Wageningen, LU.

Backerra, F., L. Flapper and A. Hobbelink (eds) (1989), *Vrouwen van het land, anderhalve eeuw plattelandsvrouwen in Nederland*. Zutphen, De Walburg Press.

Berger, B., U. Blom and T. Loeffen (1987), *Tellen agrarische vrouwen (niet) mee?* Den Haag, CVPV.

Bernet, K. (1998), *Hoe trouwen agrarische vrouwen? Een studie over agrarische vrouwen omtrent de beleving van hun huwelijksgoederenregime*. Wageningen, Wageningen University.

Bettenhausen-Verbey, J. and U. Blom (1983), *Plattelandsvrouwen over de drempel*. Den Haag, NBVP.

Bloemhoff, A. and E.J.R.M.Wolf (1987), *Arbeid en gezondheid van vrouwen werkzaam in de agrarische sector*. Voorburg, DGA.

Blom, U., C.J.M. Spierings and G.M.J. Loeffen (1986), *Boerinnen en landinrichting*. Den Haag, LEI.

Blom, U. and J.M. van de Hoek (1988), *De tijdsbesteding van boerinnen en tuindersvrouwen*. Den Haag, LEI.

Blom, U. and J.M. van de Hoek (1989), *Zwangerschapsverlof voor boerinnen en tuindersvrouwen*. Den Haag, LEI.

Blom, U. and H. Hillebrand (1992), *Jonge vrouwen op agrarische gezinsbedrijven*. Den Haag, LEI-DLO.

Bock, B.B. (1994), Women and the future of Umbrian agriculture. In: L. van der Plas and M. Fonte (eds), *Rural gender studies in Europe*. Assen, van Gorcum, pp. 91–107.

Bock, B.B. (1996), *Nieuwe onderneemsters op het platteland*, Dordrecht.

Bock, B.B. (1997), *Pluriactiviteit, vrouwen en vernieuwing*. Wageningen, Wetenschapswinkel.

Bock, B.B. (1998), *Vrouwen en vernieuwing van landbouw en platteland; de kloof tussen praktijk en beleid in Nederland en andere Europese landen*. Wageningen, Circle for Rural European Studies/Wetenschapswinkel.

Bock, B.B. (1999), *Women in rural development in Europe – appreciated but undervalued*; paper for the conference 'Gender and rural Transformations in Europe.' Wageningen.

Bock, B.B. (2000), Toenemende zakelijkheid tegenover de belangen van het moderne gezin; strategisch handelen in het veranderende gezinsbedrijf. *SPIL*, no. 167-168, 10-16.

Bock, B.B. (2001), *The problems, prospects and promises of female employment in the rural areas of Europe*, key note lecture for the conference 'The new challenge of women's role in rural Europe,' Cyprus, 4-6 october 2001.

Bock, B.B. (2002), *Tegelijkertijd en tussendoor; gender, plattelandsontwikkeling en interactief beleid*. Wageningen, WUR/ Studies van Landbouw en Platteland 32 (diss).

Bock, B.B. and M. van der Burg (1998), Vrouwenarbeid in de agrarische sector. In: H. Pott-Buter and K. Tijdens (eds), *Vrouwen; leven en werken in de twintigste eeuw*. Amsterdam, Amsterdam University Press, pp. 219-240.

Bock, B.B., P.A. Welboren and A. Lindenbergh (2000), *Emancipatie-effectrapportage stimuleringskader*. Ede, Expertisecentrum LNV) nr. 225.

Bock, B.B. and S. de Rooij (2000), *Social exclusion of small-holders and women small-holders in Dutch agriculture*. Final national report for the EU-project: Causes and mechanisms of social exclusion of women small-holders. Wageningen, Wageningen University.

Bos, M.J. (1969), *Het meewerken door vrouwelijke gezinsleden op het agrarische bedrijf*. Wageningen, Landbouwhogeschool.

Bosch, J.A. (1970), *Het meewerken van de boerin in Friesland*. Wageningen, Landbouwhogeschool.

Bosch, D. (1990), *Boerinnen maken beleid: een handleiding voor het boerinnenwerk Gelderland*. Wageningen, Landbouwuniversiteit.

Brand, M. van den and E. Verhagen (1991), *Boerinnen, belangen en belangenbehartiging*. Wageningen, Landbouwuniversiteit.

Brinke-van Hengstum, A.G.M. ten and M.Y. Toonen (1988), *Vrouwenemancipatie in landelijke gebieden*. Wageningen, Landbouwuniversiteit.

Brunt, E. (1975), Vrouwen op het platteland. *Sociologische Gids*, 22, 4.

Bun, C., A. Oldenkamp and A. Oldenziel (1990), *De agrarische vrouw en haar gezondheid in de EG*. Wageningen, Landbouwuniversiteit.

Burg, M. van der (1988), *Een half miljoen boerinnen in de klas: landbouwhuishoudonderwijs vanaf 1909*. Heerlen, De Voorstad.

Burg, M. van der (2002), *Geen tweede boer? Gender, landbouwmodernisering en plattelandsontwikkeling in Nederland 1863-1968*. Wageningen, Wageningen University, PhD thesis.

Burg, M. van der and K. Lievaart (1998), *Drie generaties in schort en overall. Terugblik op een eeuw vrouwenarbeid in de landbouw*. Wageningen, Landbouwuniversiteit/AKB.

Casteren, M. van (1997), *Meer vrouwen in besturen*. Den Haag, LTO.

Doorne-Huiskes, J. van and M. Rosendahl (1998), *Voorlichting aan agrarische vrouwen; een onderzoek naar de voorlichtingsbehoeften van agrarische vrouwen*. Utrecht.

Dorresteijn, J. (1983), *De participatie van jonge boerinnen in belangengroepen en beleidsorganisaties binnen de landbouw*. Utrecht, NAJK.

Drooglever Fortuijn, J., W. Ostendorf and F. Thissen (1994), *Vrouwen op het Friese platteland*. Amsterdam, Universiteit van Amsterdam.

Giesen, C.W.M. (1991), *Werkverhoudingen en stress op het boerenbedrijf*. Utrecht, RUU.

Groot, E.J.M. de (1989), *Boerinnen in beeld; veranderende opvattingen over boerinnen na de Tweede Wereldoorlog*. Wageningen, Landbouwuniversiteit/Vakgroep Huishoudkunde.

Haan, H.J. de (1993), Images of family farming in the Netherlands. *Sociologia Ruralis* 33 (2), pp. 147-166.

Haan, H.J. de and A.T.J. Nooij (1985), Rural sociology in the Netherlands. Developments in the seventies. *The Netherlands' Journal of Sociology* 21 (1), pp. 51–62.

Haenen, G. (1992), *Toekomst van het agrarische vrouwenwerk; integratie van agrarische vrouwen in de Noord-Brabantse Christelijke Boerenbond.* Wageningen, Landbouwuniversiteit.

Harpe, C. and M. Potters (1991), *Bedrijfsopvolgsters: belemmeringen en participatie in agrarische jongerenorganisaties.* Wageningen, Landbouwuniversiteit.

Heel, A. van and T. Lodder (1994), *Vrouwen trekken hun spoor: agrarische vrouwen op de bres voor de land- en tuinbouw.* Utrecht, CVPV.

Hendrikse, A. and L. Klaver (1995a) *Kaas in de badkuip.* Wageningen, Landbouwuniversiteit/NBVP.

Hendrikse, A. and L. Klaver (1995b) *Agrarische vrouwen vernieuwen het platteland.* Wageningen, Landbouwuniversiteit.

Hilhorst, T. (1993), *Agrarische vrouwen benoemen hun belangen.* Wageningen, Landbouwuniversiteit.

Hillebrand, H. and U. Blom (1993), Young women on Dutch family farms. *Sociologia Ruralis,* 33, 2, pp. 178–189.

Hobbelink, A. (1982), *Je trouwt niet alleen met een boer ... maar ook met het bedrijf, zijn familie en de hele buur*t. Nijmegen, KUN.

Hobbelink, A. (1991), Boerinnengeschiedenis 1940–1990. *Spil,* 93/94, pp. 5–10.

Hobbelink, A. and S. Spijkers (1986a) De mooie kamer beter benutten: boerinnenstudies in Nederland. *LOVER,* 13, 1, pp. 47–50.

Hobbelink, A. and S. Spijkers (1986b) Naar meer zicht op de verhoudingen tussen de seksen in de landbouw. *Spil,* 53/54, pp. 4–50.

Hoek, C. van de, L. Klaver and A. de Moor (1987), *Agrarische vrouwen en landinrichting.* Wageningen, Landbouwuniversiteit.

Hoerchner, J.H., J.A.C. Ophem and E. van Oorschot (1979), *Situatie van de boerin in de lidstaten der Europese Gemeenschappen.* Wageningen, Landbouwhogeschool.

Jansen, H. (1990), *Boerinnen van 50 tot 70: een verkennend onderzoek naar de gevolgen van bedrijfsopvolging, bedrijfsbeëindiging en ouder worden.* Nijmegen, KUN/LUW.

Jansen, H. (1991), *Kleine baas of grote knecht? Een verkenning van de arbeid van agrarische vrouwen vanuit structuratie-theoretisch perspectief.* Nijmegen, KUN.

Jansen, M., T. Lodder and G. Overbeek (1985), *Geen rek maar plek, landbouwbeleid voor vrouwen.* Wageningen, Boerinnengroep.

Janssen, U. and R. Lammerts (1999), *Leefbaarheid op het platteland. Sociale en culturele ontwikkelingen.* Utrecht, Verwey-Jonker Instituut.

Keuzenkamp, S. (1997), *Emancipatie-effectrapportage Herijking van de landinrichting.* Nijmegen, KUN.

Klaver, L. and J. van Poppel (1994), *Boer en boerin.Een proces van bewustwording en kiezen.* Wageningen, Landbouwuniversiteit.

Klaver, L. and S. Zwart (1996), *De complexe verhouding tussen familie en bedrijf in de landbouw; nieuwe lijnen voor toekomstig onderzoek.* Wageningen, Landbouwuniversiteit.

Klein Douwel, J. and M. Toonen (1986), *AAW voor boerinnen en tuindersvrouwen: een slechte regeling vol met hindernissen.* Wageningen, Landbouwuniversiteit.

Kloprogge, J.J.A. (1973), *Friese boerinnen over beroepsverandering.* Den Haag, LEI.

Kools, Q.H. (1994), *Emancipatiebeleid en -activiteiten van agrarische opleidingscentra.* Wageningen, Landbouwuniversiteit.

Kools, Q.H. (1996), *Agrarisch onderwijs, ook voor mesks en wichten! Een onderzoek naar vrouwelijke leerlingen in het voorbereidend en middelbaar agrarisch onderwijs.* Wageningen.

Kooy, G. (1959), *De zelfkazerij van Midden-Nederland: een onderzoek naar haar voortbestaan.* Assen, Van Gorcum.

Kriellaars, F.W.J. (1951), *Enige beschouwingen over het gezinsbedrijf in de landbouw*, Leiden, Stenfert Kroese.

Laval, W. and T. de Rijk (1990), *Meisjes, MAS en arbeidsmarkt.* Amsterdam VSLPC.

Lieshout, D. van (1989), *Bevallen en opstaan: ook de agrarische vrouwen verdient een zwangerschaps- en bevallingsverlofregeling.* Wageningen, Hogeschool Diedenoort.

Litjens, M., T. Loeffen and C. Valkhoff (1979),*Achter gaat voor.*Wageningen,Landbouwhogeschool.

Litjens, M. and L. Klaver (1991), *Hoe kan het ook anders: een onderzoek naar belangen van agrarische vrouwen.* Den Haag, NBVP.

Lodder, T. (1991), Voor wie gewoon boerin wil blijven: een beroepsstatus voor agrarische vrouwen. *Spil*, 93/94, pp. 27–32.

Lodder, T. and N. Tims (1988), *Tussen vernieuwing en traditie: het emancipatieproject herwaardering van arbeid in de Nederlandse Bond van Plattelandsvrouwen.* Wageningen, Landbouwuniversiteit.

Loeffen, G.M.J. (1984), *Boerinnen en tuindersvrouwen in Nederland.* Den Haag, LEI-DLO.

Lohuis, B.A.M., J.W.C. Maas and S.M.van der Meulen (1987), *Posities van vrouwen en meisjes op het platteland.* Wageningen, Vakgroep Huishoudkunde.

Melita, F. (1993), *Sociale zekerheid in de landbouw in vijf Europese landen. Over de pensioen en arbeidsongeschiktheidsregelingen voor boer(innen) en in vergelijking tot die voor werkne(e)m(st)ers; met speciale aandacht voor de positie van de boerin.* Wageningen, Landbouwuniversiteit.

Mok, A.L. (1990), Kwaliteit van de arbeid en het arbeidsproces. *Sociologische Gids*, 37, 3.

Nederlands Agrarisch Jongeren Kontakt (NAJK) (2000), *De bonte WAZ. Ervaringen van agrarische vrouwen met de zwangerschaps- en bevallingsuitkeringsregeling binnen de Wet Arbeidsongeschiktheidsverzekering Zelfstandigen.* Utrecht, NAJK Vrouwenoverleg & LTO commissie Vrouw en Bedrijf.

NIROV (1990), *Vrouwen in landelijk gebieden: emancipatie binnen bereik.* Den Haag, NIROV.

Nooij, A. (1997), Modern and endogenous development. Civilization and empowerment. In: H. de Haan and N. Long (eds), *Images and Realities of Rural Life.* Assen, Van Gorcum, pp. 107–120.

Oomens, M. and G. Kriekaard (1989), *De problematiek van het onproblematische: een verkennend onderzoek naar de leefbaarheid van kleine kernen en hun omringende omgeving bekeken vanuit de positie, het standpunt en de ervaringen van vrouwen*, Tilburg, PON.

Overbeek, G. (1993), Man/vrouw–maatschap beslist goed? *Tijdschrift voor Arbeidsvraagstukken*, 9, 4, pp.384–39.

Overbeek G. (1995), Agrarische vrouwen op zoek naar een baan buitenshuis? *Spil*, 131–132/133–134, pp. 49–54.

Overbeek, G. (2000), *Verzilveren of besparen. Arbeids- en vermogenssituatie van vrouwelijke zelfstandigen in de landbouw en verblijfsrecreatie.* Den Haag, LEI.

Overbeek, G. (2001), Individualisering loont? Inkomen en vermogen van vrouwelijke zelfstandigen in de land- en tuinbouw en verblijfsrecreatie. In *TSL* 16 (1), pp. 24–33.

Overbeek, G., S. Efstratoglou, M.S. Haugen and E. Saraceno (1998), *Labour situation and strategies of farm women in diversified rural areas of Europe*, Luxembourg, EC.

Pijnenburg, A. (1991), *De toekomst van de Katholieke Plattelandsvrouwen Nederland: een onderzoek naar perspectieven van een organisatie voor plattelandsvrouwen.* Wageningen, Landbouwuniversiteit.

Ploeg, J.D. van der (1999), *De virtuele boer*. Assen, Van Gorcum.

Ploeg, J.D. van der, H. Renting and M. Minderhoud-Jones (eds) (2000), The socio-economic impact of rural development: realities and potentials. *Sociologia Ruralis*, 40, 4, (special issue).

Ploeg J.D., A. van Cooten, T. Kierkels, A. Logeman (eds) (2002), *Kleurrijk platteland; zicht op een nieuwe land- en tuinbouw*. Assen, Van Gorcum.

Poppel, J. van, A. Dees, T. Lodder and N. Raedts (1990), *Stapsgewijs, wijs-worden: integratie van agrarische vrouwen in de standsorganisaties*. Den Haag, CVPV.

Rooij, S. de (1992), *Werk van het tweede soort. Boerinnen in de melkveehouderij*. Assen/ Maastricht, Van Gorcum.

Rooij, S. de (1994), Work of the second order. In: M. van der Burg and M. Endeveld (eds), *Women on family farms. Gender research, EC policies and new perspectives*. Wageningen, Agricultural University) pp. 67–78.

Rooij, S. de (1998), Boerinnen en tuindersvrouwen – vrouwenpower op het platteland. *Facta*, 6, 8, pp. 12–15.

Rooij, S. de, E. Brouwer and R. van Broekhuizen (1995), *Agrarische vrouwen en bedrijfsontwikkeling*. Wageningen, Landbouwuiversiteit/WLTO.

Saal, C.D. (1958), *Het boerengezin in Nederland: sociologische grondslagen van gezin en bedrijf*. Assen, Van Gorcum.

Sinnema, G. (1990), *Kiezen MAS-meisjes anders? Onderzoek naar het keuzeproces van MAS-leerlingen*. Wageningen, Landbouwuniversiteit.

Spierings, C.J.M. (1972), *Boeren en boerinnen over beroepsverandering*. Den Haag, LEI.

Spierings, C.J.M. (1974), *Jonge boerinnen over beroepsverandering*. Den Haag, LEI.

Stork-van der Kuyl, D.M.E.J. (1952), *De Drentse boerin. Haar plaats in de samenleving*. Assen, van Gorcum.

Stork-van der Kuyl, D.M.E.J. (1966), *De boerin van nu en straks: vrouwen in een veranderende plattelandssamenleving*. Assen, van Gorcum.

Sweere, A. (1991), *Zowel boer als boerin: vrouwelijke ondernemers in de land- en tuinbouw*. Wageningen, Boerinnengroep.

Verboon, H. (1988), Huishoudelijke voorlichting, in: H.P. de Bruin, (ed.), *Het Gelderse rivierengebied uit zijn isolement; een halve eeuw plattelandsvernieuwing*. Zutphen, De Walburg Pers, pp. 192–199.

Wal, D. van der (1990), *Bijscholing agrarische vrouwen: verslag van een onderzoek onder meewerkende agrarische vrouwen in de provincie Utrecht naar leerbehoeften voor deelname aan agrarische cursussen*. Wageningen, Landbouwuniversiteit.

Wijk A. van (1983), Wat de boer niet ziet, dat telt hij niet: meitellingen laten veel arbeid verborgen. *Landbouwmaand*, 12, 2.

Wijk, A. van (1986), *Boerinnengroepen werken zo: over jonge agrarische vrouwen en boerinnengroepen in Gelderland*. Wageningen, Landbouwhogeschool.

Wijk, I. van (1996), *Zij staat haar mannetje. Onderzoek naar de identiteit en het imago van agrarische vrouwen in bestuurlijke functies*. Wageningen, Hogeschool Diedenoort.

Zwart, S.I. (1990), *Agrarische huishoudens: een onderzoek naar de veranderende relatie tussen gezin en bedrijf in Oost-Gelderland*. Wageningen, Landbouwuniversiteit.

Zwart, S.I. (1996), *Boerinnen en bezit*. Paper voor de Sociaal-Wetenschappelijke Studiedagen 11–12 april 1996, Amsterdam.

Zwieten, D. van (1994), *Agrarische gezinnen in zorgen.Een inventariserend onderzoek naar sociaal-psychische en sociaal-economische problemen van agrarische gezinnen*. Wageningen, Landbouwuniversiteit.

Chapter 11

Discourses on Rurality and Gender in Norwegian Rural Studies

Nina Gunnerud Berg

The aim of this chapter is to outline some of the main connections between rural studies and gender studies in Norway and to illustrate them with empirical work. Do the development of the two fields have points of resemblance or have they developed differently? How are the two linked together in the joint field 'Feminist rural studies'? What characterizes feminist rural studies? What has motivated and what have been the aims of the research? What questions have been asked? What theoretical and methodological approaches have been adopted? These are the questions I want to examine in this chapter. The period considered stretches from the mid-1970s, when the first work on women and rural development was conducted, until today. The review is not a comprehensive treatment of the history of feminist rural studies. My aim is rather to give a brief summary of the changing emphases over the period by focusing on the ways particular debates in rural studies and gender studies have influenced feminist rural studies. Thus, I will try to situate feminist rural studies within the development of gender and rural studies, so as to explain changes in approach to work on rurality and gender.

Rural Studies and Gender Studies in Norway

In Norway research on rural development has been conducted under the rubric of 'Regional development studies' as well as 'Rural studies.' Regional development studies have traditionally focused mainly on rural areas, meaning all parts of Norway except the urban agglomerations around Oslo, Bergen, Trondheim and Stavanger.[1] Consequently, I use the term 'Rural studies' to include 'Regional development studies.'

Rural studies, which has been a research field within Norwegian social sciences since around 1960, is conducted within university departments (geography, sociology, social anthropology, political science, and economics), regional research institutes, the Norwegian Agricultural University and, not least, within the Centre for

Rural Research in Trondheim (established 1982) which is the biggest single rural research environment. Gender studies in Norway dates back to the beginning of the 1970s when 'Women's studies' was added to the social sciences as a minority subject. Today gender perspectives are acknowledged as a legitimate area for social scientific inquiry, but gender issues still remain marginal to most social scientists. Like rural studies, gender studies is conducted both within university departments with distinct disciplinary affiliations and within multidisciplinary centres. Centres for Women's studies/gender studies are located in all the four university towns in Norway (Oslo, Bergen, Trondheim and Tromsø). In fact, both fields are characterized by multidisciplinarity. A second similarity is that the key concepts – gender and rurality – have been conceptualized more or less the same way during the 25 year period. Thirdly, and to some extent related to conceptualizations of gender and rurality, power relations is a key dimension in both fields. Last, both are highly influenced by Anglo-Saxon research as regards conceptual and theoretical approaches taken, rural studies more so than gender studies. Actually, Norwegian gender research has delivered important theoretical contributions. These have been developed 'bottom-up,' since originally the studies were empirical-political in character (Widerberg, 1992, 1997).

It is the connections between rural studies and gender studies as coming together in the joint field 'Feminist rural studies' that is the main focus of the rest of this chapter. I want to illustrate how the focus within the field has shifted from subordinate women in depressed rural areas to emphasize questions about the construction of masculinities, femininities and ruralities. But before doing so, I want to underline that feminist rural studies is, as both its mother-fields, a multidisciplinary field. Research projects are often conducted by a group of researchers educated within different social sciences who share an interest in rural areas and gender relations. They often publish together or one author reports from a multi-disciplinary project. Consequently, I do not focus on what different disciplines have contributed to the field, nor to the referred authors' disciplinary backgrounds.

Categories of 'Feminist Rural Studies'

A relatively rich literature on gender and rural development has developed over the past 25 years. I have, therefore, found it necessary to categorize this work and look for the main characteristics of each category. Although I identify categories of work that have developed chronologically, this division in phases is not unproblematic. Competing perspectives have been held at the same time and contemporary research is diverse. The categories therefore have permeable boundaries and are unstable entities. Despite this, I think that a chronological classificatory framework helps in understanding the changing emphases within the field. Consequently, the categories are to some extent pedagogical conveniences.

I have attempted a classification that mainly reflects perspectives and theoretical approach, and have come to the conclusion that feminist rural studies in Norway since the mid-1970s fall into three categories in which different dimensions of gender and rurality are the focus of attention:

1. Politically motivated (Women's Liberation Movement) research more or less influenced by women's studies focusing on rural women's subordination.
2. Research founded on feminist theories and concepts focusing on gender relations in rural areas.
3. Research founded on theories and concepts within rural studies and gender studies (and some attempts at integrating the two) focusing on masculinities, femininities and ruralities.

I have labelled the categories 'Rural women's subordination' (1), 'Gender relations in rural areas' (2) and 'Masculinities, femininities and ruralities' (3), respectively. The three-fold categorization forms the main structure of the rest of the chapter in which each category is treated in separate sections.

Rural Women's Subordination

The category of 'Rural women's subordination' is dominated by research from the second half of the 1970s and the early 1980s, and the majority of the work referred to in this section was conducted in that period. However, much research conducted more recently falls into this group.

Aims

The overall aim of mainstream rural studies in Norway in the 1960s and the early 1970s was to make the rural areas visible. Much research on rural areas was motivated by a wish to document inequalities and unfairness relative to urban areas and argue for a regional policy directed towards rural areas, especially remote rural areas, so that they could catch up with the development in urban areas. Rural areas were, in other words, represented as depressed, traditional areas that did not fit the norm, i.e. the modern urban areas. Both rural research and policy was, therefore, based on a rural–urban dichotomy and rural and urban areas were defined functionally using statistical indicators, mainly on economic activities. Rural–urban divisions of labour were very much focused upon.

As regards gender studies, or to be more correct, women's studies in Norway in the 1970s and early 1980s, these were strongly motivated by the Women's Liberation Movement. The main aim of the research was to make women visible. Women's studies were at the time defined as 'studies by, on and for women' (Holter, 1996, my translation). The strategy chosen was to illustrate the unfairness of women's inequalities relative to men and argue for equal rights, so that women could catch up with men, i.e. fit the male norm (equal rights feminism). The works were, therefore, based on a women–men dichotomy, and gender roles was the key concept used to 'explain' the inequalities. Socio-sexual divisions of labour was the predominant area of investigation.

It was within this context of early mainstream rural studies and gender studies that the field 'Feminist rural studies' was born in the second half of the 1970s. At the same time there was a political climate in which especially rural area problems, but increasingly also women's rights, were recognized as important. During the early 1970s Norway had its one and only period of what might be called 'counter-urbanization,' i.e. a decline in migration towards the biggest cities and an increase in the migration towards rural areas mainly caused by an enormous growth of the public sector in rural areas and helped by the 'green wave.' As the building up of the public sector in rural areas was gradually fulfilled and migration flows turned around again in the second half of the 1970s, it became urgent to find new measures to stabilize the settlement pattern. The fact that young women migrated from rural to urban areas to an even greater extent than young men called for empirical research that could give the reasons and suggest means to reduce such a tendency.

Consequently, the overall aim of the first feminist rural studies was to explain women's flight from rural areas. A dominant hypothesis was that the explanation was to be found in women's subordinate position within traditional rural economic activities and in rural communities in general. The researchers were, therefore, eager to describe the socio-sexual divisions of labour in rural areas, in productive as well as reproductive work, in order to document rural women's subordination. In short, the goal was to make women in rural areas visible.

Questions

Women's subordinate position in agriculture, is a dominant theme within this category of work. Farm women and social security issues were among the first themes focused on (Almås, 1977), as was the historical development of women's work within agriculture (Almås, Vik and Ødegård, 1983; Almås, 1983). Research looked critically at labour relations on the family farm from the perspective of women's roles and responsibilities. The concept 'masculinization of agricultural work' was adopted to explain how modernization processes, not least mechanization, had led to a situation in which men had gradually taken over more and more of the farm work (Blekesaune and Almås, 1991; Almås and Haugen, 1991; Blekesaune, 1993). That was not to say that farm women were out of work. On the contrary, it was documented that farm women have many different 'helping out tasks' in addition to housework and child care, and that their everyday lives are characterized by long working hours often resulting in health problems. However, they do not have corresponding shares in decision making and income (Wiggen, 1976; Vik, 1984; Haugen, 1985; Fyhn, 1985; Borchgrevink,

1989; Dale, 1991, 1993; Bolstad, 1993). In short, the farm woman's role as a flexible, supportive spouse helping the farmer was made visible.

Women's everyday lives in coastal areas is a related theme also focused very much on within 'Rural women's subordination' (Gerrard, 1976, 1982; Bratrein, 1976; Lie, 1976; Solberg, 1976; Larsen, 1980; Flakstad,1982; Holtedahl, 1982). In common with the studies on farm women, the studies document subordination and hard work on the part of women, both employees in the fishing industries and those living on farms. Fisher-farmer holdings – historically the mainstay of North Norwegian coastal communities (Jones, 2000) – involved that farming was the main responsibility of women. Men were away fishing for a significant part of the year. Larsen (1980) stresses that the term 'fisher-farmer' understood as a man combining two occupations, should be replaced by the concept 'care-farmer', understood as a woman combining care (for children and/or elder relatives, cooking and all kind of housework) and farming on a daily basis. The man in the fisher-farmer household was mainly a fisher who took part in the farm work for short periods. The term 'care-farmer' is one example of theoretical contributions developed bottom-up in early feminist studies.

Works on rural women's education and employment possibilities within others than traditional rural economic activities (Nyseth, 1983, 1985; Jevnaker, 1985) and women entrepreneurs (Sandvig, 1985) are also present within this category, as are works on women and rural policy (Fredriksen, 1981, 1982; Valestrand, 1984a; Teigen, 1984).

Theoretical and Methodological Approaches

The 'Rural women's subordination work' is, in accordance with the rural studies and the women's studies influencing them, first and foremost characterized by dichotomic thinking. As mentioned above, rural–urban is the main dichotomy within rural studies, women–men is the equivalent in women's studies. The dichotomic perspective in which urban and men are the superior or the norm, combined with an emancipatory aim in which equality between rural and urban areas and between women and men is the final goal, explain the empirical studies in which documentation of the subordination of rural women is the main goal. Rurality and gender are mainly conceptualized as variables. Quantitative methodologies characterize the studies that first and foremost aim at documentation, while qualitative methodologies are adopted in studies that seek to deepen the descriptions by help of non-quantifiable aspects. In fact this goes also for some of the earliest studies, as women's studies realized the shortcomings of quantitative methodologies earlier than social sciences in general. In other words, making visible rural women's lives often includes making visible how women themselves experience their rural lives.

The earliest studies within this category no doubt firmly placed gender on the agenda of rural studies and formed the basis for later rural studies aiming at a better understanding of the gendering of rural societies.

Gender Relations in Rural Areas

Work in this category date back to around 1985 and most of the publications referred to in this paragraph were produced in the late 1980s and the early 1990s. The category does, however, comprise even more recent contributions.

Aims

Norwegian rural studies from the late 1970s and the early 1980s are characterized by a different perspective than earlier work in that the descriptions of the miserable countryside are replaced with an optimism on behalf of the rural areas. The possibilities and advantages of the countryside are often focused upon. The urban as a norm is questioned and a valorization of the rural is part of a 'green wave' found in politics as well as research. The legitimate definition of 'the rural' is still functional, but socio-cultural relations are added to economic activities as indicators. Rural areas are increasingly represented as idyllic, not least as places of community, i.e. areas with close human relationships, mutual cooperation and support. The idea of stabilizing the settlement pattern entered Norwegian regional policy in the beginning of the 1970s (during the above mentioned period of counter-urbanization) and has permeated it ever since. The strategies chosen in the 1980s can be summed up in the concept of 'Rural development programmes' in which local empowerment was a central dimension. Rural researchers participated partly in developing the programmes, partly in evaluating them.

The primary difference between the women's studies that inspire work in the 'Rural women's subordination' category and the gender studies that influence rather strongly the work in this second category, 'Gender relations in rural areas,' is the way in which gender difference is theorized. What first and foremost characterizes Norwegian gender studies from the early 1980s is its valorization of gender differences. Instead of seeing gender differences as unfair, work in this perspective celebrates difference and attempts to reverse the traditional allocation of superiority to all that is masculine. It is, in other words, questioning the male as norm. The idea of gender-specific identities, rationalities, perspectives and standpoints (Sørensen, 1982; Wærness, 1984) plays a central role (standpoint feminism). The movement away from construing gender issues in the language of equality towards that of difference – and sometimes female superiority – is mirrored in arguments for women's participation programmes in which women's empowerment is a central aspect. The concept of 'gender relations' gradually replaced 'gender roles' thus underlining that gender is a relative concept and that women should not be considered as victims, but rather as agents negotiating the meaning of gender.

The 'Gender relations in rural areas studies' are influenced by the changed emphases in rural studies and gender studies in the early 1980s. The descriptive approach of the 'Rural women's subordination studies' was then partly replaced by, and partly supplemented with an approach aimed at understanding the gendering of rural life. The studies in the category are, furthermore, influenced by a political climate in which concerns about rural women's situation had gradually resulted in a will to initiate rural development programs aimed at women. For example, the problematic situation of farm women was so centrally placed on the political agenda that in 1987 the Norwegian Research Council initiated a five-year research program on 'Women in Agriculture'.

Questions

The title of the program ('Women in Agriculture') reflects a coming change in conceptualization of women's contributions in agriculture, both in politics and research. The term 'farm women' commonly used earlier to include all women living on farms, from those who live on a farm but do not take part in the farm work to those who manage a farm on their own, was replaced by a whole range of terms for sub-groups, such as for example, 'farm housewife,' 'working farm wife,' and not forgetting: 'Women/female farmers' – a term that highlighted that there are women who are the main operators on farms. It was, in fact, launched as early as 1982 by Fyhn (Fyhn, 1982), but it was not until the late 1980s that it became more widely accepted and came into use. Haugen (1990a, 1994a) focuses on questions such as: Who are the women farmers, how do they become farmers and how do they farm? What is achieved when women enter into male domains? Is equality between the sexes achieved, and are women's values being expressed in the way women farm?[2] Within the broader theme 'women's way of farming' we find a lot of sub-themes, as for example female farmers' use of technology, not least the tractor (Brandth and Bolsø, 1991; Brandth, 1991; Bolsø, 1991).

As women gained equal rights to become successors on farms in 1974 through an amendment to the law regulating succession rights to farms, it was the time to study the effects of the law as the 1980s turned into the 1990s. Not least, it was important to gain insights into whether young girls want to run farms, why they do or do not, what type of opportunities and barriers they expect to face, or have faced, and what attitudes towards women farmers their parents and grandparents have (Ølnes, 1989; Haugen, 1993, 1994b; Bjørnstad, 1997; Ahlin, 1999).

Other roles of women living on farms also investigated within the 'Gender relations in rural areas category' included the role of the entrepreneur as initiatives to broaden the economic basis of the farm became necessary (Brattested, 1989; Bolsø, 1987), the daughter-in-law (Bye, 1999), the participant in agricultural organizations (Meistad, 1993), and the 'diet and health expert' (Eriksen, 1988) – just to mention a few. Furthermore, works on farm women in an historical perspective should not be forgotten (Thorsen, 1993; Fossgard, 1996).

Like the studies on gender relations in agriculture, studies on gender relations in coastal areas with fishing-related economic activities aim at conceptual investment. For example, Gerrard's (1990) critique of the concept 'fisher woman/wife' parallels the critique of the concept 'farm woman/wife.' Furthermore, the studies parallel the studies on women in agriculture in that they aim at valorization and try to shed light on how women negotiate the gender roles of the traditional fisher-farmer household. Although in the process of becoming history since the 1960s, research shows that women along the coast experience the roles as ascribed to them, and partly adopt, and partly resist them (Flakstad, 1984; Valestrand, 1984b, Holtedahl, 1986; Gerrard, 1994, 1999; Pettersen, 1997; Grønbech, 1999; Kramvig, 1999; Elstad, 1999). Furthermore, gender relations in the fishing industry and the factory ship fleet are investigated (Larsen and Munk-Madsen, 1990; Husmo and Munk-Madsen, 1994; Jentoft, Thiessen and Davis, 1994; Husmo, 1994; Munk-Madsen, 1994) as is politics of fisheries (Lotherington and Thomassen, 1994; Gerrard, 1995).

In addition to being studied as a way to broaden the economic basis of farms entrepreneurship is examined as a potential strategy for women in general to have a job and an income in rural areas. Studies aim at explaining why there are so few women entrepreneurs by shedding light on the difficulties women face when trying to set up and run a firm as well as focusing on women's way of handling entrepreneurship. Women's entrepreneurship is presented as different from men's. The keywords used to describe women's way of becoming and being entrepreneurs are 'ethic of care', 'rationality of responsibility', 'family-orientation', etc. (Berg, 1991, 1994; Alsos and Ljunggren, 1995; Ljunggren, Magnussen and Pettersen, 2000).

While earlier feminist rural studies explained women's 'flight' from the countryside by women's subordinate position in traditional rural economic activities, the focus was gradually broadened to include a whole range of aspects influencing women's everyday life in rural and urban areas respectively. While traditional gender relations in the countryside are presented as push factors, education and work often are presented as pull factors in the cities, (see for example, Wiborg, 1990a,b; Wold, 1994; Dahlstrøm, 1996; Grimsrud, 2000).

Theoretical and Methodological Approaches

The work in the 'Gender relations in rural areas' category differs in rather fundamental ways from the work in the 'Rural women's subordination category'. Valorization of women and rural areas replaced victimization. It was realized that the one-sided account of women and rural areas as miserable fails to see women and rural areas as a resource – a consequence of the equality perspective that characterizes the studies. In the 'Gender relations in rural areas' category the equality perspective is replaced by a difference perspective in which the superiority of women and rural areas is advocated. This is connected to the understanding of rurality and gender as standpoints, rather than variables. Such a conceptualization is in turn related to the aim of understanding instead of describing. Although many areas of investigation are the same in the two categories, the rather one-sided focus on economic aspects (rural–urban and gendered division of labour) within the 'Rural women's subordination category' are broadened to include social aspects (the qualities of rural and urban areas, gendered identities, etc.). It is not least the focus on women's ways of doing things and the tendency to explain their ways with a 'rationality of responsibility' (and men's ways with a technical-economic rationality) that characterize the work. The dichotomic thinking colouring the works in the 'Rural women's subordination category' is, in other words, also present in this category in that gender and rurality are seen as central analytical categories.

It should be noted, however, that both Norwegian rural studies and gender studies throughout their histories have been sensitive to differences. Rural studies' sensitivity to differences among rural areas has, no doubt, to do with the 'geography' of Norway. In a very long and narrow country comprising different climates, landscapes, cultures, etc. the differences between – to take but one example – a fishing community in North Norway and a farming community in the South East are too significant to be overlooked. As early as in the 1970s gender researchers problematized talking about 'the woman' and generalizing about women (and men) as undifferentiated categories (see Haavind, 1976). Although the studies in a standpoint perspective first and foremost are characterized by focus on and discussion of the rural women specific, considerations of differences among rural women dependent on region or place are taken into account. In other words, the basic idea behind the feminist rural studies, namely that rural women live different lives from urban women and that their experiences and practices as well as rural life in general need to be analysed, is combined with a sensitivity to differences among rural women.

Qualitative methodologies, not least open-ended interviewing, dominate the research in this category. This is, of course, closely connected to the conceptualization of rurality and gender as 'standpoints' and the aim of understanding the significance of gender in different situations and contexts. A basic understanding of gender relations as closely connected to concepts as identity and rationality – and consequently concerning things that can not or should not be measured or weighed – lies behind the choice of interviews as the dominant technique for data collection.

The third and last category of feminist rural studies comprises work from the late 1990s onwards. This will be discussed in the next paragraph.

Masculinities, Femininities and Ruralities

Aims

In rural studies since the late 1980s the categories 'rural areas' and 'urban areas' have been increasingly questioned since the differences between the country and city are becoming blurred. Generalizations about rural areas have more or less been replaced by a focus on diversity. This is reflected in an increasing use of the plural form: 'ruralities.' Deconstruction of the established category is in line with an approach to rurality as socially and culturally constructed, and places analyses of representations of rurality at the centre stage. The overall aim is to identify and unpack existing representations and explore the significance and consequences of them. An understanding of rurality as a social construct is not as evident in political discourses as much as in academic discourses, but there is a change concerning how rural areas are conceived of as policy target areas. 'Improving the image of rural areas' has recently been suggested as a policy measure (KRD, 2000).

In gender studies from the last ten to fifteen years one can observe an increasing awareness of the changing and fluid nature of masculinity and femininity. The categories 'masculinity' and 'femininity' are, in other words, questioned and are replaced by 'masculinities' and 'femininities' to indicate plurality. Gender is conceived of as a social and cultural construct that varies in time and space. While research until around 1990 tended to look away from men as gendered individuals and focused almost exclusively on women, recent research has expanded its scope to cover also men and masculinity. In political discourse some attention has been paid to men and masculinity, but in regional development discourses men as gendered individuals are absent. For example, the recent White Paper on regional politics (St.meld nr. 34, 2000–2001) focuses on women (and youth) as groups experiencing the countryside as problematic (Berg, 2001).

Recent feminist rural studies have, in accordance with the changed emphases in rural and gender studies, taken a new turn. The overall aim of the studies in the category 'Masculinities, femininities and ruralities' is to analyse women's and men's day-to-day experiences of rurality to uncover the assumptions and expectations associated with constructions of rural gender identities. One seeks, in other words, to explore the intertwining of constructions of rurality and gender. Such work is, however, still in its infancy and only a limited number of questions are approached.

Questions

The traditional rural economic activities and entrepreneurship are issues also studied within this category of work. The masculine gendering of agriculture and forestry is examined by Brandth and Haugen (Brandth, 1994a, b; Brandth and Haugen, 1998; Haugen, 1998). In seeking to render the taken-for-granted masculine gendering of agriculture and forestry visible and identify the construction and dynamics of hegemonic masculinity they read gender into the texts of tractor advertisements (Brandth, 1993, 1995) and a forestry magazine (Brandth and Haugen, 2000). The masculine gendering of fishing (Munk-Madsen, 1996) and fish processing (Husmo, 1999) is likewise touched upon, and Hauan (1999) discusses representations of coastal masculinity. Foss and Berg (forthcoming) analyse how the gendering of entrepreneurship is related to life-course and place.

Furthermore, hunting – which is an integral part of rural life in Norway that has strong associations with masculinity – is considered. Taking as her point of departure that some women hunt, Bye (2000) examines the 1998 volumes of two Norwegian journals on hunting to analyse to what degree women hunters are a subject of comment and the way masculinity and femininity are represented in pictures and texts.

Yet another theme researched is the use and the construction of the meaning of communication technology among youths in rural areas. Lægran (in press, 2002) analyses the rural internet cafes as gendered techno spaces for youths.

Urban-rural migration and new rural households' everyday life in the country-side are themes also covered within this third category (Berg, 1998, 2002; Berg and Forsberg forthcoming). This work focuses on how constructions of rurality encompass strong expectations of the identities of men and women and of the gender relations within the home and the rural community. Furthermore, it highlights that constructions of rural masculinity and femininity are taken into account by both men and women when considering moving from the city to the countryside, since they assume that to belong to the rural community is to comply – at least more or less – with accepted notions of being a woman or a man.

Theoretical and Methodological Approaches

The two last categories ('Gender relations in rural areas' and 'Masculinities, femininities and ruralities') have in common a focus on gender relations. What makes the understanding of gender in the third category different from the one in the second is the focus on difference, diversity and change that has encouraged researchers to think beyond and deconstruct the categories 'women' and 'men.' As regards the understanding of rurality there is a marked difference between the two. In the second category rurality is understood as 'something out there,' i.e. distinct, bounded spaces. In the third category the relationship between the rural and geographical space is more unsettled. Attention has shifted to the rural as a mental category that is socially constructed, and it is recognized that there are multiple constructions. The dichotomic thinking applied to both rurality and gender in the two first categories is, in other words, questioned in this third category.

The conceptualization of rurality and gender as social and cultural constructs gradually placed analyses of social representations of rurality and gender at the centre stage of research. Unpacking of hegemonic social representations of the rural and other more or less marginal ones ('other countrysides') have begun to play a central role within rural studies, and in gender studies hegemonic social representations of masculinity and femininity are analysed. Implicit in thinking of different social representations as ranked is an awareness of the fact that power is bound up discursively in the very social and cultural constructs. Representations of rural masculinities and femininities are connected to ideas about what it is to be a real rural man or a real rural woman.

Qualitative methodologies characterize the work in this third category, and open-ended interviewing is still a dominant method of data collection. The focus on social representations of rurality and gender has, however, brought about discourse analyses. Researchers try to uncover the processes by which meanings are established, how they rather become 'natural' and how they influence social practices. These aims have encouraged the use of textual, photographic and filmic data sources.

Conclusions

Hopefully my account of feminist rural studies in Norway has shown that the historical development of rural studies and gender studies has a lot of similarities, not least as regards discourses on the key concepts, i.e. rurality and gender. I argue that there are at least three distinguishable sets of similarities between discourses on rurality and gender throughout the 25–30 years of rural studies and gender studies. Each set of conceptualization dominates each of the three categories of feminist rural studies identified above. To distinguish between them one might borrow from linguistics the categories noun, adjective and verb. This has been done earlier within gender studies to distinguish between different understandings of the concept of gender, and I think the same reasoning goes for rurality. As regards the category 'Rural women's subordination' in which rurality and gender are conceived of as variables, one might argue that they are also thought of as nouns – women, men, areas, regions, places. In the second category – Gender relations in rural areas – rurality and gender are conceived of as standpoints or characteristics, and one might say that they are thought of as adjectives, i.e. to express rurality and gender one uses adjectives – female, male, idyllic, safe, clean, beautiful, etc. In the third category – Masculinities, femininities and ruralities – in which rurality and gender are conceived of as ranked social representations, one might argue that they are thought of as verbs. In gender studies it is argued that gender is something we do – a thought first presented by West and Zimmerman as early as 1987 in the well-known article 'Doing gender' (West and Zimmerman, 1987). In rural studies the idea that different cultural expectations and practices constantly redefine rurality and that there is contrasting knowledge of how to behave in the rural setting (Halfacree and Boyle, 1998) parallel the idea that gender is something we do. We have different cultural competence with respect to what rural life is about and act according to these – we are doing rurality.

Yet another way of separating the three discourses is to stress their different perspectives, which is closely connected to different ideas about power. As regards the understanding of rurality and gender as variables or nouns as in the 'Rural women's subordination' category equality between urban and rural areas and between women and men is seen as the goal. The differences are considered undesirable. The studies have, in other words, an equality perspective and equality means equal amount of power. According to the understanding of rurality and gender as standpoints or adjectives, as in the 'Gender relations in rural areas' category, gender differences are considered desirable, and it is the differences that are focused upon. It is a perspective seeing women as different from and sometimes superior to men. It is, consequently, a women-centred perspective. The idea is that women's ways of thinking and acting should be acknowledged and that the process will empower women. In the understanding of rurality and gender as social representations or verbs as in the 'Masculinities, femininities and ruralities' category, it is plurality that is in focus. It is recognized that social representations are ranked, and that hierarchies of masculinities, femininities and ruralities can be identified. Since certain gender identities are celebrated and reinforced through the circulation of dominant meanings of rurality they influence social practices and distinguish between legitimate and illegitimate practices. The power of social representations are, in other words, acknowledged and taken into account in this perspective.

As mentioned above, the themes thus far addressed within this approach are relatively limited. Feminist rural research energies should now, I think, seek to develop the perspective further and be directed towards gender issues not yet approached as well as towards 'old' issues. By way of a final concluding paragraph, I want to highlight very briefly what I see as some key topics and directions for future research.

We still know very little about men's identities within the rural community and how these are worked out in their daily lives. We need a much more detailed understanding of the expectations and performance of male gender identities, both central and mainstream constructions of rural masculinity and, for example, gay masculinities. There remains a lack of published work on the lives of both gay men and lesbians living in rural areas. Philo's (1992) much quoted article on 'neglected rural others' in which he argues that the lives of the poor, the elderly, women, those with disabilities, black people and ethnic minorities are ignored in rural research has – with the exception of women – had little impact on Norwegian rural studies so far. Research on otherness and marginality is in short supply. Furthermore, there is a clear need for research to focus on how the body is constructed and imagined in a rural context and, for example, to analyse the relationship between the body and the use of space in the countryside. These are just a few possible research avenues for the future. Since gender is now an accepted and legitimate research interest within rural studies, I believe that fascinating works on gender and rurality will appear that will both bring new and interesting findings and generate theoretical and conceptual ideas.

Notes

1. 'Rural Norway' is commonly considered to be all parts of Norway except the four biggest urban agglomerations (see e.g. Norges Forskningsråd, 1999). This has not least to do with the fact that Norway is far more scarcely populated than countries in central and southern parts of Europe – 14,5 inhabitants per square kilometer as compared to e.g. Great Britain with 241.
2. 'Though the majority of women in Norwegian agriculture can be classified as agricultural helpers or agricultural partners in Pearson's term (Pearson, 1979), there is a group of farm women who are either without husbands or have husbands that are not involved in day-to-day farm work. Between five and ten percent of Norwegian farm operators are women. A female farmer is defined in this article as a woman who is the sole or main operator of the farm' (1990, p. 197). The article from 1990 is characterized by an equality perspective and belongs consequently to the 'subordination category', while the one from 1994 combines an equality perspective and a difference perspective.

References

Ahlin, Britt Helen (1999) *Odelsjenter i Overhalla. Ønsker odelsjenter i Overhalla å bruke odels-retten sin, og hvilke forhold spiller inn i deres valg?* Hovedoppgave i geografi, Geografisk institutt. Trondheim, NTNU.

Almås, Reidar (1977) *Norsk jordbruk – det nye hamskiftet.* Oslo, Gyldendal Norsk Forlag.

Almås, Reidar (1983) Maskulint og feminint på landsbygda i dag. SFB *notat* 3/83, Trondheim, Senter for bygdeforskning.

Almås, Reidar and Marit S. Haugen (1991) Norwegian gender roles in transition. The masculinization hypothesis in the past and in the future. *Journal of Rural Studies* 7, pp. 79–83.

Almås, Reidar, K. Vik and Jørn Ødegård (1983) Women in rural Norway. Recent tendencies in the development of the division of labour in agriculture and the participation of rural women in the labour market. SFB *paper* 1/83, Trondheim, Senter for bygdeforskning.

Alsos, Gry and Elisabet Ljunggren (1995) Kari Askeladd og de gode hjelperne? Om forholdet mellom kvinnelige etablerere og kommunale næringsetater i Lofoten. *NF-rapport* 11/95. Bodø, Nordlandsforskning.

Berg, Nina Gunnerud (1991) *Kjønnsperspektiv på entreprenørskap i distrikts-Norge.* Allforsk, Trondheim, Senter for samfunnsforskning.

Berg, Nina Gunnerud (1994) *Servicelokalisering og entreprenørskap.* Dr. polit-avhandling. Trondheim, Geografisk institutt, UNIT-AVH.

Berg, Nina Gunnerud (1998) Kjerringer og gubber mot (flytte)strømmen – hvorfor noen flytter til bygde-Norge. Upublisert notat. Revidert versjon 2001 i Arbeider fra Geografisk Institutt, Universitetet i Trondheim, nr. 35.

Berg, Nina Gunnerud (2001) Mennesker og steder – vurdering av Stortingsmelding nr. 34 (2000–2001) 'Om distrikts- og regionalpolitikken' ved hjelp av stedsteori i geografifaget. stortingsmelding nr. 34 2000-200. Innlegg på REGUT's midtveiskonferanse i Oslo 15. – 16. oktober.

Berg, Nina Gunnerud (2002) Kjønn på landet og kjønn i byen – om betydningen av kjønnete konstruksjoner av ruralitet for urban–rural migrasjon og hverdagsliv. *Kvinneforskning* 1.

Berg, Nina Gunnerud and Gunnel Forsberg (2003) Rural geography and feminist geography – discourses on rurality and gender in Britain and Scandinavia. In Simonsen, Kirsten and Jan Øhman (eds) *Förnyelse i nordisk kulturgeografi.* Aldershot, Ashgate.

Bjørnstad, S. (1997) *Odelsjentas oppdragelse. Hvordan foreldrene påvirker odelsdatteren til å gjøre et valg om å overta gården eller ikke.* Hovedoppgave NLH, Ås.

Blekesaune, Arild (1993) Structural Changes in Norwegian Agriculture 1975–1990. From family farms to one-man farms. Young Scientist Lecture on The 15th European Congress of Rural Sociology, Wageningen, The Netherlands.

Blekesaune, Arild and Reidar Almås (1991) Bondehusholdets ulike strategier for å overleve. SFB *paper* 10/91. Trondheim, Senter for bygdeforskning.

Bolstad, Tove (1993) Kårkontrakter – menns avtaler om kvinners arbeids- og livssituasjon. In Brandth, Berit and Berit Verstad (eds) *Kvinneliv i landbruket.* Landbruksforlaget. Oslo, pp. 53–78.

Bolsø, Agnes (1987) Evaluering av etablererstipend for kvinner i landbruket. SFB- *rapport* 6/87. Trondheim, Senter for bygdeforskning.

Bolsø, Agnes (1991) Kvinner, maskiner og yrkesidentitet i landbruket. SFB-*paper* 3/91. Trondheim, Senter for bygdeforskning.

Borchgrevink, Tordis (1989) *Mye arbeid – liten innflytelse?* Oslo, Landbruksforlaget.

Brandth, Berit (1991) 'Diesel-berter' og 'fjøskjerringer' – Om likestillingsforskning i landbruket. SFB-*notat* 2/91. Trondheim, Senter for bygdeforskning.

Brandth, Berit (1993) The mutual construction of tractors and gender in farmers' everyday lives. Paper to the COST A4 workshop. Trondheim October.

Brandth, Berit and Marit S. Haugen (1998) Breaking into a masculine discourse. Women and farm forestry. *Sociologia Ruralis* 38 (3), pp. 427–442.

Brandth, Berit (1994a) Teknologi og mannlighet i forandring. Gården som arena for konstruksjon av kjønn. *Sosiologisk Tidsskrift* 3, pp.185–203.

Brandth, Berit (1994b) Changing femininity: the social construction of women farmers in Norway. *Sociologia Ruralis* 34, pp. 127–149.

Brandth, Berit (1995) Rural masculinity in transition. Gender images in tractor advertisements. *Journal of Rural Studies* 11 (2), pp. 123–133.

Brandth, Berit and Agnes Bolsø (1991) 'New' women farmers and their use of technology. *SFB-paper 7/91*. Trondheim, Centre for Rural Research.

Brandth, Berit and Marit S. Haugen (2000) From lumberjack to business manager: masculinity in the Norwegian forestry press. *Journal of Rural Studies* 16 (2000), pp. 343–355.

Bratrein, Håvard Dahl (1976) Det tradisjonelle kjønnsrollemønsteret i Nord-Norge. In Forfatterkollektiv: *Drivandes kvinnfolk. Om kvinner, lønn og arbeid.* Tromsø, Universitetsforlaget, pp.21–38.

Brattested, Astrid (1989) Kvinnelige etablerere i bygde-Norge. *SFB rapport* 2/89. Trondheim, Senter for bygdeforskning.

Bye, Linda Marie (1999) *Landbrukskvinners hverdagsliv – en studie av kjønnet arbeidsdeling og tette boforhold.* Hovedoppgave. Trondheim, Geografisk institutt, NTNU.

Bye, Linda Marie (2000) Femininitet og maskulinitet i storviltjakta. Paper no 1/00. Trondheim, Senter for bygdeforskning.

Dahlstrøm, Margareta (1996) Young women in a male Periphery. Experiences from the Scandinavian North. *Journal of Rural Studies* Vol. 12 (3), pp. 259–271.

Dale, Kristin (1991) Kvinnebønder og trygd: Fortjener kvinnene i jordbruket mer inntekt? *Norsk Landbruk* 11.

Dale, Kristin (1993) Bondekoner – jordbrukets arme riddere? In Brandth, Berit and Berit Verstad (eds) *Kvinneliv i landbruket.* Oslo Landbruksforlaget, pp. 79–106.

Elstad, Åsa (1999) Arbeidsroller i nord-norske fiskarbondehushald. In Gerrard, Siri and Randi Rønning Balsvik (ed.) Globale kyster. Liv i endring – kjønn i spenning. *Kvinnforsks skriftserie* 1/99. Tromsø, Senter for kvinneforskning og kvinner i forskning.

Eriksen, Sissel (1988) Kvinna tar ansvaret. *Nytt om kvinneforskning* 4/88, pp. 60–66.

Flakstad, Anne Grethe (1982) Kvinnevilkår i kombinasjonsbruk. *Ottar* nr. 6.

Flakstad, Anne Grethe (1984) Kan endring innen kvinnearbeidet utløse strukturendringer? In Rudie, Ingrid (ed.) *Myk start – hard landing.* Oslo, Universitetsforlaget, pp. 64–83.

Foss, Lene and Nina Gunnerud Berg (forthcoming) 'Det er her jeg skal være...' Betydningen av sted og livsløp for etablerere. Kapittel 8 i Berg, Nina Gunnerud and Lene Foss (eds), *Kjønnsperspektiver på entreprenørskap.* Oslo, Abstrakt Forlag.

Fossgard, Eldbjørg (1996) *Frå lagnad til val. Kvinneliv på vestnorske gardsbruk 1930-1990.* Jærmuseet, Nærbø.

Fredriksen, Sissel (1981) *Kjønnspolitikk og distriktsutbygging.* Oslo, Universitetsforslaget.

Fredriksen, Sissel (1982) Blir åttiårene enda et nytt mannstiår i distrikts-utbygginge? *Plan og Arbeid,* Nr. 6, pp. 412–414.

Fyhn, Asbjørg (1982) Bondekvinner eller kvinnebønder? *Ottar* nr. 6.

Fyhn, Asbjørg (1985) *Makt og kjærlighet i jordbruket. Ei drøfting av betingelsene for kvinnerekruttering til jordbrukets produksjon og organisasjon.* Hovedoppgave, Institutt for samfunnsvitenskap, Universitetet i Tromsø.

Gerrard, Siri (1976) Noen sider ved husmorarbeidet i et nordnorsk fiskevær. In Forfatterkollektiv: *Drivandes kvinnfolk. Om kvinner, lønn og arbeid.* Tromsø, Universitetsforlaget, pp. 56–67.

Gerrard, Siri (1982) Kvinnefellesskap i et fiskevær. In Harriet Holter (ed.) *Kvinner i fellesskap,* Oslo, Universitetsforlaget, pp. 213–230.

Gerrard, Siri (1990) Fiskerkona som begrep og fenomen i norsk samfunnsvitenskapelig fiskeriforskning. *NIBR-notat* 1990: 137. Oslo, Norsk institutt for by- og regionforskning.

Gerrard, Siri (1994) Kvinners forvaltning – havets ressurser. In Otterstad, Oddmund and Svein Jentoft (eds) *Leve kysten.* Oslo, Ad Notam, pp.123–134.

Gerrard, Siri (1995) When women take the lead: Changing conditions for women's activities, roles and knowledge in North Norwegian fishing communities. *Social Science Information* 34/4.

Gerrard, Siri (1999) Festivaler og bygdedager som lokale stemmer. Eksempler på kvinners og menns lokale aktiviteter i et fiskekvoteregime. In Gerrard, Siri and Randi Rønning Balsvik (eds) Globale kyster. Liv i endring – kjønn i spenning. *Kvinnforsks skriftserie* 1/99. Tromsø, Senter for kvinneforskning og kvinner i forskning, pp. 168–179.

Grimsrud, Gro Marit (2000) Kvinner på flyttefot. *ØF-rapport* 13/2000. Lillehammer, Østlandsforskning.

Grønbech, Dagrunn (1999) Kystkvinnen – som omsorgsbonde og ressursforvalter. In Gerrard, Siri and Randi Rønning Balsvik (eds) Globale kyster. Liv i endring – kjønn i spenning. *Kvinnforsks skriftserie* 1/99. Tromsø, Senter for kvinneforskning og kvinner i forskning, pp. 47–62.

Halfacree, Keith and Paul Boyle (1998) Migration, rurality and postproductivist countryside. In Boyle, Paul and Keith Halfacree (eds) *Migration into rural areas. Theories and issues.* John Wiley and Sons, pp.1–20.

Hauan, Marit Anne (1999) 'Velkommen, du kjører no inn i Harry-land'. Kystens menn i urbanitetens diskurs. In Gerrard, Siri and Randi Rønning Balsvik (eds) Globale kyster. Liv i endring – kjønn i spenning. *Kvinnforsks skriftsserie* 2/99, pp. 157–167.

Haugen, Marit S. (1985) Bondekvinners arbeid og helse. *SFB rapport* 5/85. Trondheim, Senter for bygdeforskning.

Haugen, Marit S. (1990a) Female Farmers in Norwegian Agriculture. From Traditional Farm Women to Professional Female Farmers. *Sociologia Ruralis* 30 (2), pp. 197–209.

Haugen, Marit S. (1990b) Kvinnebonden. *SFB-rapport* 7/90. Trondheim, Senter for bygdeforskning.

Haugen, Marit S. (1993) Odelsjenter – likestilte i loven, men hva med praksis? In Brandth, Berit and Berit Verstad (eds) *Kvinneliv i landbruket.* Oslo, Landbruksforlaget, pp. 23–52.

Haugen, Marit S. (1994a) Gender differences in modern agriculture. The case of women farmers in Norway. *Gender and Society,* 8 (2) pp. 206–229.

Haugen, Marit S. (1994b) Rural women's status in farming and property law; lessons from Norway. In Whatmore, Sarah, Terry Marsden and Phillip Lowe (eds) *Gender and Rurality.* Critical Perspectives on Rural Change Series VI. London, David Fulton Publishers Ltd., pp. 87–101.

Haugen, Marit S. (1998) The gendering of farming. The case of Norway. *The European Journal of Women's Studies* 5, pp. 133–153.

Holtedahl, Lisbeth (1982) Lokalkultur er kvindekultur i Veggefjord. In Harriet Holter (eds) *Kvinner i fellesskap.* Oslo, Universitetsforlaget, pp. 198–212.

Holtedahl, Lisbeth (1986) *Hva mutter gjør er alltid viktig.* Oslo, Universitetsforlaget.

Holter, Harriet (1996) Kvinneforskning, Utvikling og tilnærminger. In Holter, Harriet (ed.) *Hun og han. Kjønn i forskning og politikk.* Oslo, Pax Forlag, pp. 58–77.

Husmo, Marit (1994) Kvinnelige fiskeindustriledere. In Otterstad, Oddmund and Svein Jentoft (eds) *Leve kysten.* Oslo, Ad Notam, pp. 191–200.

Husmo, Marit (1999) Menn definerer – kvinner praktiserer. Om kvalitetssikring i norsk fiskeindustri. In Gerrard, Siri and Randi Rønning Balsvik (eds) Globale kyster. Liv i endring – kjønn i spenning. *Kvinnforsks skriftsserie* 1, pp.105–122.

Husmo, Marit and Eva Munk-Madsen (1994) Kjønn som kvalifikasjon i fiskeindustrien. In Otterstad, Oddmund and Svein Jentoft (eds) *Leve kysten.* Oslo, Ad Notam, pp. 179–190.

Haavind, Hanne (1976) Det kvinner har felles er kjønnsmessig undertrykking. In Forfatterkollektiv, *Drivandes kvinnfolk.* Om kvinner, lønn og arbeid. Tromsø, Universitetsforlaget, pp. 112–132.

Jentoft, Svein, Victor Thiessen and Anthony Davis (1994) I samme båt? Fiskerkvinner som bakkemannskap. In Otterstad, Oddmund og Svein Jentoft (eds) *Leve kysten.* Oslo, Ad Notam, pp. 145–150.

Jevnaker, Birgit Helene (1985) Jente i bygda. *Plan og Arbeid* 2/85, pp. 40–49.

Jones, Michael (2000) Resources and Representations: Sea and Highland as the 'Two Landscapes' of North Norway. Paper presented at the Permanent European Conference for the Study of the Rural Landscape. Aberystwyrth, 10th–17th September.

Kramvig, Britt (1999) Ære og verdighet. Kvinnelighet og mannlighet i et fiskerisamfunn. In Gerrard, Siri and Randi Rønning Balsvik (eds) Globale kyster. Liv i endring – kjønn i spenning. *Kvinnforsks skriftserie* 1/99. Tromsø, Senter for kvinneforskning og kvinner i forskning, pp. 63–77.

KRD (2000) Distrikts-Norge – Hvor ligger mulighetene? Hva må gjøres? http.//odin.dep.no/krd/

Larsen, Marit and Eva Munk-Madsen (1990) Kjønnsmyter med konsekvenser. En analyse av skillet mellom kvinner og menn i industriell fiskeforedling til lands og til vanns. *NFFR-rapport*, Norges Fiskerihøyskole, Univeristetet i Tromsø.

Larsen, Sidsel Saugestad (1980) Omsorgsbonden – et tidsnyttingsperspektiv på yrkeskombinasjon, arbeidsdeling og sosial endring. *Tidsskrift for samfunnsforskning* 21 (3–4), pp. 283–296.

Lie, Merete (1976) Kvinner i fiskeindustrien. In Forfatterkollektiv: *Drivandes kvinnfolk.* Om kvinner, lønn og arbeid. Tromsø, Universitetsforlaget, pp. 76–87.

Ljunggren, Elisabet, Tone Magnussen and Liv Toril Pettersen (2000) Bedriftsetablerere mellom hushold og marked. In Husmo, Marit and Jahn Petter Johnsen (eds) *Fra bygd og fjord til kafébord?* Trondheim: Tapir Akademiske Forlag, pp. 121–135.

Lotherington, Ann Therese and Bente Thomassen (1994) Organisering av kystkvinnene. In Otterstad, Oddmund and Svein Jentoft (eds) *Leve kysten.* Oslo, Ad Notam, pp.135–144.

Lægran, Anne Sofie (2002) The petrol station and the internet cafe: Rural techno spaces for youth. *Journal of Rural Studies.*

Lægran, Anne Sofie (forthcoming) Escape vehicles? The Internet and the Automobile in a local/global intersection.

Meistad, Torill (1993) Kvinne og tillitsvalgt i meierisamvirket – møte med en mannskultur. In Brandth, Berit and Berit Verstad (eds) *Kvinneliv i landbruket.* Oslo, Landbruksforlaget, pp. 131–149.

Munk-Madsen, Eva (1994) Moderskabets møde med fabriksskibet. In Otterstad, Oddmund and Svein Jentoft (eds) *Leve kysten.* Oslo, Ad Notam, pp. 159–168.

Munk-Madsen, Eva (1996) *Fiskerkøn.* Avhandling, Norges Firskerihøgskole, Universitetet i Tromsø.

Norges Forskningsråd/The Norwegian Research Council (1999) Regional utvikling Oslo.

Nyseth, Torill (1983) Har de unge jenter noen plass i Finnmark? *Ottar* nr. 3.

Nyseth, Torill (1985) Stengt for ungjenter. Jentenes situasjon i Finnmark. *Plan og Arbeid* 2/85, pp.19–24.

Pettersen, Torill (1997) *Hovedsaken er at kjerringa er i arbeid. Husholdsstrategier i fiskerikrisen.* Hovedoppgave i sosiologi. Trondheim NTNU.

Philo, Chris (1992) Neglected rural geographies: a review. *Journal of Rural Studies*, 8, pp.193–207.

Sandvig, Anne Margrete (1985) Etablererskolen for kvinner: Kvinner kan, vil og våger. *Plan og Arbeid* 2/85, pp. 58–65.

Solberg, Anne (1976) Jente- og guttearbeid. In Forfatterkollektiv: *Drivandes kvinnfolk.* Om kvinner, lønn og arbeid. Tromsø, Universitetsforlaget, pp. 68–75.

Solheim, Liv (1984) Ein skal vere sterk for å klare seg her. Om familieflytting og kvinnetilpassing i eit lokalsamfunn. In Rudie, Ingrid (ed.) *Myk start – hard landing.* Oslo, Universitetsforlaget, pp. 301–317.

St.melding nr. 34 (2000–2001) Om distrikts- og regionalpolitikken. Oslo.

Sørensen, Bjørg Aase (1982) Ansvarsrasjonalitet: Om mål-middeltenkning blant kvinner. In Holter, Harriet (ed.) *Kvinner i fellesskap*, Oslo, Universitetsforlaget, pp. 392–402.

Teigen, Håvard (1984) *Vil Bygde-Norge overleve – med storbukkar på bygdebeite?* Ringsaker, Fagbokforlaget.

Thorsen, Liv Emma (1993) *Det fleksible kjønn. Mentalitetsendringer i tre generasjoner bondekvinner, 1920–1985.* Oslo, Universitetsforlaget.

Valestrand, Halldis (1984a) Likestillings-, distrikts- og næringspolitikk, – inkommensurable størrelser? *Arbeidsrapport nr. 4, ISV*, Universitetet i Tromsø.

Valestrand, Halldis (1984b) Fra fiskerkone og fabrikk-kvinne til forstadshusmor. In Ingrid Rudie (ed.) *Myk start – hard landing.* Oslo, Universitetesforlaget, pp. 103–122.

Vik, K. (1984) Bondekvinnens medbestemmelse i gårdsdrifta. In Almås, Reidar, Sissel Eriksen and Martin Rønningen (eds) *Leve Bygdenorge.* Trondheim Senter for bygdeforskning, pp. 86–94.

West, C. and Zimmerman, D.H. (1987) Doing gender. *Gender and Society* 1, pp. 125–51.

Wiborg, Agnete (1990a) Jenters utdannelse – enveisbillett ut av distriktene? *Rapport 3/90, Arb eidsforskningsinstituttet*, Bodø.

Wiborg, Agnete (1990b) Lokalsamfunnet i utkanten; startsted, mellomstasjon eller endestopp. Unge kvinners tilpasning i distriktene. *Rapport 4/90, Arbeidsforskning*, Bodø.

Widerberg, Karin (1992) Teoretisk verktøykasse – angrepsmåter og metoder. In Taksdal, Arnhild og Karin Widerberg (eds) *Forståelser av kjønn i samfunnsvitenskapenes fag og kvinneforskning.* Oslo, Ad Notam Gyldendal, pp. 285–299.

Widerberg, Karin (1997) Norsk kvinneforskning – i ett 'outsider-within' -perspektiv. *Kvinneforskning* 3–4 pp. 56–60.

Wiggen, Asbjørg (1976) Kvinner i jordbruket. In Forfatterkollektiv: *Drivandes kvinnfolk.* Om kvinner, lønn og arbeid. Tromsø, Universitetsforlaget, pp. 39–55.

Wold, Anne Grethe (1994) *Kjønns- og stedsidentitet – av betydning for 'kvinneflukten' fra bygde-Norge?* Hovedoppgave Geografisk institutt, Universitetet i Trondheim.

Wærness, Kari (1984) The rationality of caring. *Economics and Industrial Democracy* 5, pp. 185–211.

Ølnes, Anna (1989) Odelsjenter – barrierer og muligheter. *SFB-rapport* 1989. Trondheim, Senter for bygdeforskning.

Chapter 12

Rural Women in Development Discourses and Social Science Research in Tunisia

Alia Gana

This chapter analyses the ways in which women and gender issues have been dealt with in relationship to rural development, both by policy makers and social scientists. Identifying the changing problematic within which rural gender issues have been inserted, according to shifts in development strategies, the chapter reviews the major themes on which social science research and national policies related to women and rural development have focused. This overview covers the period that followed Tunisia's independence in 1956, but gives particular attention to the ways in which gender issues are reframed within the context of liberalization processes and rural restructuring.

Modernization and Rural Women: From the 1960s to the mid-1970s

At the time of independence, the major part of the Tunisian population (60 percent) lived in rural areas. Poverty, limited access to land resources and illiteracy were the main characteristics of the rural population. In the 1960s, agricultural modernization through the implementation of a co-operative-based land reform has been viewed as the most effective way to promote access of the rural population to economic and social progress. Viewed as a fortress of traditionalism and archaism, the peasantry was considered as a possible threat to the modernization project. In the 1960s and early 1970s, policy makers and social scientists both expressed this idea. Structural reforms in the countryside and the ways in which peasants could hinder or foster the modernization process have in fact particularly attracted the attention of social scientists (mostly geographers and sociologists) and have resulted in a very abundant literature, published mainly in the *Revue Tunisienne des Sciences Sociales* (Zghal, 1969; Ben Salem, 1993).

During this period, however, both policy makers and researchers devoted no direct attention to rural women. Only to the extent that rural women were considered as having a possible influence (negative or positive) on the development process were they the target of development policies or a subject of interest for researchers.

In fact, the status of rural women, in particular their lack of education (92.1 percent of rural women were illiterate in 1966) and their subordinated position within the patriarchal household, was viewed as a possible threat to the modernization project. Women issues were thus mainly approached from the perspective of women's role and influence on the rural family. How to reinforce their positive influence on the family and therefore on the development process was the main objective of policy makers as far as rural women were concerned. In this context, both education and family planning were assigned a major role in the modernization of the rural family. Through education and a change of mentality, rural women would play an important role in the achievement of development goals (reduction of population growth, participation in the modernization of the rural economy and society, reduction of the rural out migration).

Launched in the early 1960s, the *family planning programme* (import of and advertising for anti-conceptional products were authorized in 1960 and abortion became illegal in 1965) reflected the modernist approach on which Tunisia's development policy was based. Mainly directed towards rural women, the programme was assigned a major role in the reduction of population growth in rural areas. Also by contributing to the transformation of the status of rural women within the household, the programme was likely to play a role in the weakening of traditional social forces and consequently in the reinforcement of state power in rural areas. Several authors (Bchir, 1976; Boukraa, 1976) have analysed family planning and its consequences on the status of rural women as a powerful tool for the consolidation of the modern national state. Together with family planning, women's education was also viewed as having a major influence on the modernization project.

Since independence, education has constituted a fundamental option of Tunisia's development strategy and it was considered as an important instrument in the struggle against 'obscurantism' in rural areas. The Tunisian government has constantly promoted universal primary education (education is free of charge and mandatory until the age of sixteen), which was considered as a major factor of emancipation and progress, and as a crucial instrument in ensuring national cohesion and stability. In addition to national policies aiming at increasing the level of education of the Tunisian population and at improving school enrolment rates of girls, special programmes in favour of rural women and girls have been implemented since the first decade after independence. Among these programmes the *Centres de la Jeune Fille Rurale* (learning centres for rural girls), implemented in rural areas and managed by UNFT (National Union for Tunisian Women), were assigned an important role in sustaining state efforts towards the modernization of the rural family and society.

Target groups of these learning centres are illiterate rural girls between the age of 12 and 18. In the 1960s and 1970s, literacy tuition, domestic education, nutrition and sanitary education were the main components of the centres' teaching programmes. Education objectives were mainly oriented towards preparing rural girls to better achieve their role of spouses and mothers, following a modern model. Education programmes were also aimed at promoting positive images of the land and the countryside among rural women so that they could contribute to reducing rural out-migration. Whereas literacy and domestic education occupied the largest place in national programmes targeting rural women in the 1960s and early 1970s, professional education did not retain as much attention, although we should mention the creation in the 1960s of two agricultural training centres where young girls with primary education are trained in farming techniques, in addition to classes in domestic, sanitary and civic education. However, until recently, training programmes for rural girls have remained very much oriented towards preparing them for their role of mothers and farmer's spouses. In 1965 the *Ecole Nationale de Monitrices Rurales de la Soukra* was created. Its mission was to train young girls with secondary education to become agricultural extensionists, instructors in the centres de *Jeunes Filles Rurales*, or farm women. Very rarely however, graduates of the women's agricultural school have been able to use their diploma to practice an autonomous farm activity, access to land being very limited for rural girls and women.

Hence during the 1960s and early 1970s, rural women's contribution to social progress, when taken into account in development policies, was mainly viewed from the perspective of their role in the family. Women's work and contribution to economic activities in rural areas were given little attention, mainly because these activities were viewed as pertaining to the traditional sector and, for that reason, rural women were not considered as potential actors in the economic development process in rural areas. Although in the social science literature some authors started to be interested in women's work as a new social phenomenon in Tunisia, the focus was almost exclusively on urban women and wage work (Baffoun, 1981; Hotchschlid, 1967).

Rural Women, Food Security and Rural Development: From the mid-1970s to the late 1980s

As stressed earlier, the ways in which rural women's issues have been approached since Tunisia's independence have varied with shifts in development models and strategies, both at national and international levels. National policies and social science research on rural women have been particularly influenced by development approaches advocated by international organizations (FAO, UN, World Bank).

From the mid-1970s to the mid-1980s the context within which rural women's issues are raised is characterized by increasing food deficits and growing disparities between urban and rural areas. During this period national policies are thus increasingly oriented towards ensuring food self-sufficiency and include a more important attention to rural development. This context promotes a rising interest and an increasing awareness of women's role in the rural and agricultural economy, which is expressed both in development programmes and in social science research, and bears two main aspects: 1. A recognition of the roles played by women in rural households' food security and the implementation of programmes aiming at enhancing women's role in domestic production. 2. A growing attention to female farm work and women's contribution to market-oriented farm production. In addition, rural women's issues start to be taken into consideration in rural development approaches and programmes.

Rural Women and Regional Inequality

Starting in the mid-1970s, growing regional inequalities and income disparities between the rural and the urban population lead to the implementation of the Rural Development Programme. Within this framework, rural women issues are taken into account mainly with regard to two main aspects: improvement of living conditions (housing, access to potable water) and consolidation of family-based production (productive family programme, family gardens projects). As stressed by the former president of the Tunisian Women's Organization at that time (Mzali, 1981), rural development polices should aim at improving living conditions in the countryside and at valorizing rural life and farm work among rural women, so that they can contribute to maintaining their families in the countryside and to discouraging rural out-migration.

During that period, however, rural development programmes did not consider women as a specific target group. In general, and more specifically in social science research, women were hardly ever associated with rural and regional development, except for some rare articles (Karoui, 1989).

Rural Women and Domestic Production

The second issue in relation to which rural women are considered is that of domestic production. On the development side, small projects aiming at sustaining women's role in subsistence production, especially in poor households, start to be implemented both by governmental and non-governmental organizations (Ministry of social affairs, Women's Union and rural NGOs). The scope of these programmes (productive family programme, family gardens projects) has remained very limited, however, as well as their impact on the living conditions of poor families. It should be noted in this regard that the conception of these programmes has continued to reflect the ways in which rural women were viewed at that time, that is, mainly as subsistence producers.

In the late 1970s, rural women's role in domestic production becomes also a subject of interest for social scientists. During this period, Ferchiou, a Tunisian anthropologist, published a whole series of articles dealing with this subject. In her studies of rural communities of southern Tunisia (1978 a,b,c, 1979), she is more specifically interested in analysing the contradiction between the importance of female contribution to the rural household economy and the fact that women are not recognized as producers. She explains this contradiction by the persistence of the patriarchal family system within which women occupy a dominated position. Her approach to rural women's economic role closely links the household sexual division of labour to a critical analysis of the traditional patriarchal system of social relations (1978a, 1978b).

Women's Work, Food Security and Market Production

In the early 1980s a progressive shift of attention from women's role in subsistence production to women's work, particularly in agriculture, is to observe. A pioneer work by Sophie Ferchiou (1980, 1985), carried out in the framework of a development project in the Sidi Bouzid region and funded by a Swedish co-operation, inaugurates a whole series of studies devoted to the description and analysis of female activities and the sexual division of labour in farm production and the rural economy (Ben Younes, 1984, 1987; Gana, 1989). These studies conducted at local level in various regions of Tunisia are often linked to development projects and funded by bilateral co-operation (Germany, Sweden) or international organizations. They particularly flourished in the early 1990s. Based on their results a typology of women's farm work and contribution to the rural economy was made possible (Rivière, 1987; Gana, 1988). These studies have contributed significantly to making women's work in agriculture and their contribution to the rural economy more visible.

The focus of the studies carried out on rural and farm women during the 1980s varied according to their sponsors and the analytical approaches of the researchers. Studies initiated by development organizations were mainly preoccupied with how to better integrate rural women in the development process and to increase their productivity (in farm production). In 1988, the World Bank sponsored a study on the integration of women in agricultural development (Gana, 1988), which led to the elaboration of an agricultural extension programme for rural women. The Ministry of Agriculture, with the support of FAO, then implemented this programme whose objective was to improve rural women's skills in farming techniques.

In addition to being raised in relationship to food security problems, rural women's issues started also to be raised in relationship to the question of incomes in rural areas. How to achieve women's contribution to income generation in rural households has become also a new concern for development planners, especially since the late 1980s. This approach of rural women's economic integration would be expressed in agricultural and rural development projects implemented during the 1990s, particularly in their 'income generating activities' components targeting women.

In addition to making women's work more visible, studies carried out in relationship to development projects have contributed to spread a less negative image of rural women, which focuses more on the hardship of their work and their life rather than on their ignorance and traditionalism. Some authors (Ferchiou, 1985) insisted however on women's exploitation and on their dominated status within the patriarchal family, which tends be reinforced as a result of the development of the market economy.

Whereas most studies on women's work in urban areas have been devoted to wage work, this is not the case for rural women. Women's wage work in agriculture did not attract the attention of researchers, with the exception of Karoui's study on *Les ouvrières agricoles de la région de Mateur* (1980), devoted to female work in co-operative farms. It should be mentioned that, until the mid-1980s, women represented only a small minority of farm wage workers, especially of those employed on a permanent basis. As we will see in the section below, female wage work in agriculture has considerably increased in the most recent period and is starting to attract more attention of researchers.

Hence between the late 1970s and the late 1980s, a new subject of interest, both for development planners and social scientists has emerged: that of rural women's work and economic integration. Education and family planning have continued however to be important areas of concern for developers and researchers, but they have also started to be thought of more importantly in relationship to women's economic integration. As far as vocational training is concerned, an article by Zghal (1986) stresses that these training programmes were not well adapted to the needs of rural girls and that they were not based on the valorization of existing rural activities and local knowledge. On the contrary, they tended to promote an urban type of female activities that increased rural women's dependency on urban markets and ended up having little impact in terms of their economic integration.

Rural Women and Economic Restructuring

It is especially during the 1990s that rural gender studies have flourished in Tunisia. Increasing interest in women's issues in rural areas has to be explained in relationship to the new context brought about by structural adjustment policies implemented in Tunisia since the late 1980s. In fact, processes of economic restructuring and shifts in state policies have resulted in major transformations in the organization of agricultural production and patterns of rural livelihoods. Cuts in state subsidies, agricultural price liberalization, and the reorganization of the farm credit system, while reallocating resources in favour of large corporate farms, significantly altered the economic environment of family farming, particularly its access to land resources. While fostering the reorganization of agricultural production along more capital-intensive lines and significantly altering the conditions of social reproduction in agriculture, new forms of Tunisia's integration in the world economy have resulted also in important movements of restructuring and job cuts in the industry. In this context of growing demand imposed on agriculture, but also of increasing social costs of economic restructuring, state policies have aimed, on the one side, at improving technical skills of agricultural workers, including women and, on the other side, at promoting the diversification of rural households' activities and income sources, particularly of those based on family and female labour. Within this framework, rural gender studies since the early 1990s have been concerned with the following four areas: work and economic contribution, education and vocational training, poverty alleviation and the promotion of self-employment and, women and participatory rural development.

Work and Economic Contribution

Interest in rural women's work and economic contribution already started in the late 1980s. Most studies were initiated in relationship to national development programmes and projects funded by co-operation agencies (FAO, World Bank, IFAD, and German co-operation). While mainly aimed at the identification of development programmes in favour of rural women (technical training and promotion of income generating activities), these studies often included an assessment of women's farm work (gender division of labour, time use) and of their contribution to rural households' income. In addition to studies based on the analysis of national level data and statistics, and stressing the importance of rural women's contribution to the national economy (Makhlouf, 1993; Triki, 1995; Gana, 1991, 1996), a number of monographs, conducted at the local level and often linked to development projects, aimed at assessing the role of rural women in farm production and the rural household economy.

Most of these studies (both macro- and micro-level types) remained very descriptive. While making women's farm work more visible, some pointed however to the feminization of agricultural labour (Triki, Gana). Available data (Ministry of Agriculture, 1994–95) reveal in fact that women represent 46 percent of the total agricultural labour force and 64 percent of the family labour force employed in the farm sector. Women represent also an increasing share of agricultural wage labour, especially of seasonal workers (38 percent). According to the authors, this process of feminization of farm labour results from of a combination of factors, which include male out-migration and pluriactivity, but also the development of an intensive and more diversified farm sector (irrigated farming and fruit and vegetable production), relying mainly on cheap labour.

Studies on farm women carried out in the 1990s include also a critique of statistical data and categories, which do not allow for an adequate apprehension of female labour in agriculture (Triki, 1995; Gana, 1999). These studies call for the need to improve official statistics and quantitative approaches of farm labour, so that they can better account for women's work. Taking into account this demand to improve gender statistics, a number of reports have been ordered, namely by the Ministry of Agriculture, with the aim to assess the contribution of female labour to agricultural production and to improve methodological tools for collecting statistical data on women's work in agriculture (Kamoun, 1994; Drira, 2000).

Critique of farm labour statistics has also led to the implementation of time-use surveys of farm- and rural women. In her study on 'Time-use of rural women', Triki (2000) shows how women in the North-West region of Tunisia are overburdened with work and play a crucial role in the households' income generation.

Other authors (Gana, 1999) point however to the reductionist character of analyses of rural women's activities, which are only based on statistical data and time-use surveys, and call also for the development of more qualitative approaches to female labour. For example, analyses based on statistical data and categories cannot account for the actual role of women in the decision-making process at farm household level. If available statistics (Ministry of agriculture, 1994–95) report only a small percentage (5.6 percent) of women heads of farm enterprises, this is far from reflecting the reality. In fact, a number of women who have actually a primary role in running the family farm (the husband having migrated to urban areas or having an off-farm job) is not counted in official statistics as farmers. Several authors insist on the necessity to focus more on conceptual issues, to move beyond descriptive and statistical accounts of farm women's work and to incorporate analyses of female activities in the larger context of the changing social and economic systems within which the latter are inserted.

Studies of the 1990s contributed a great deal to increasing the awareness of the importance of women's role in farm and rural activities and of the necessity to better integrate women in development policies.

Education and Vocational Training

Despite important progress made since Tunisia's independence in the area of women's education, gaps are still to observe between rural and urban areas and illiteracy rates among rural women have remained high. In the recent period, this situation has attracted the attention of women's organizations and policy makers. Social scientists have responded to this demand for a better identification of the factors hindering an improvement of the level of education of rural women. In 1992, UNFT (National Union of Tunisian Women) ordered a study on 'School dropouts among rural girls', which showed that a variety of factors contributed to girls' school enrolment rates in rural area. These factors include long distances from houses to schools, the obligation for rural girls to take charge of domestic tasks or to take employment to help the parents. The study suggests a number of actions likely to diminish school dropouts among rural girls. Its conclusions were used by the Ministry of education and led to the implementation of an UNICEF funded programme aimed at improving school enrolment and success rates in rural areas.

In addition to concerns expressed about persistent rural/urban and men/women inequalities in the area of education and about their negative consequences on economic development, increasing attention to women's participation in vocational training programmes starts to develop during the 1990s. In fact, vocational training for rural women comes back to the forefront, but this time in a slightly different way. While in the 1960s and the 1970s, training and education were mainly viewed as a way to reinforce the positive role of women in the rural family, the focus is now on how vocational training could strengthen women's productive role.

In the early 1990s, efforts aimed at better integrating rural and farm women in development policies consisted mainly in the implementation of agricultural extension programmes in favour of rural women. A number of studies on rural women have been carried out in relationship to agricultural development projects with the aim to assess women's technical training needs. On the basis of these surveys, female agricultural extension programmes have been designed and implemented in different regions. However, as highlighted by several reports and impacts studies (Gana, 1996) most of these extension programmes have tended to confine rural women in their 'traditional' activities (small animal rearing, transformation of fruits and vegetables, handicraft etc.) rather than to better integrate them into the agricultural development process.

The failure of most agricultural extension programmes in favour of farm women contributed in the mid-1990s to a reorientation of female vocational training approaches in rural areas (GTZ, 1993, Schurings, 1999). Instead of farm women, the main targets of the new vocational training programmes are rural girls. Executed by the Ministry of Vocational training, their main objective is to contribute to a better insertion of rural girls (with basic education) into the labour market. Hence, the focus is more on how to diminish female unemployment in rural areas and less on how to increase the productivity of female farm labour. Also vocational training for rural girls is increasingly viewed as a way to promote self-employment and the development of female small and micro-enterprises in rural areas and as an element of the diversification of the rural economy (instead of being only linked to agricultural production as this used to be earlier).

Within the context of these changing development approaches and strategies, social scientists have been often asked to contribute to help decision makers and practitioners to better identify their actions and the needs of the targeted population, i.e., rural women. During the 1990s, they have been increasingly involved in social assessment studies, project evaluation and identification of development components in favour of rural women.

Poverty Alleviation and the Promotion of Self-Employment

As mentioned earlier, movements of economic restructuring have profoundly altered patterns of rural livelihoods and have resulted in increased rural poverty. Within this context, rural women issues have been increasingly dealt with in relationship to poverty problems and poverty alleviation policies and programmes. Drawing on new approaches developed by international organizations and some aspects of the literature on gender (feminization of poverty), a number of studies have been devoted to women's status and living conditions in poor areas of rural Tunisia (Bchir, 1993; Mahfoudh, 1993). Often linked to rural development projects, these studies are geared towards the identification of income generation activities for poor rural women. One common characteristic of these studies is that they remain very descriptive and rarely integrate their analyses of women's living and working conditions within the larger socio-economic context, which prevails at regional and national level. Therefore they often fail to identify appropriate programmes and actions in favour of the different social groups that rural women belong to. Development programmes identified on the basis of these studies are uniform, mainly consisting in small activities, such as vegetable gardens, small animal rearing, and handicraft, and are often unable to provide rural women with regular incomes.

The promotion of female self-employment as poverty alleviation intervention has most recently led the Tunisian government to place particular emphasis on supporting the development of micro-enterprises. Within this framework a number of development institutions and NGOs have started to implement micro-credit programmes in favour of rural women in different regions of Tunisia. In relation to these programmes, a few studies aimed at assessing the impacts of micro-credit on the socio-economic status of rural women have been conducted. Studying the experience carried out by a rural development NGO in a disadvantaged area of Northern Tunisia, Gana (2000) analyses the dynamics generated by the implementation of micro-credit programmes, drawing particular attention to the commoditization process of the household economy brought about by these programmes. Female micro-projects contribute in fact to increasing household cash income and generate growing monetary needs. They also have major impacts on the organization of household production and family labour. The author points to the changing status of women's work in rural households, which tends to bear a productive character, and contributes to individualize women's participation in the household economy. It appears however that economic dynamics generated by micro-projects rarely allow for the consolidation of an autonomous and sustainable micro-enterprise. Activities promoted by women remain fundamentally dependent on the resources available on the family farm, and in particular on their access to land, which continues to be very limited.

In recent years, micro-credit programmes in favour of women have developed and have been implemented in most regions of rural Tunisia. They are executed by development NGOs. Conceived as a way to promote self-employment and to provide rural households with additional incomes, their actual impacts on the economic status of women have not sufficiently attracted the attention of social scientists.

Another growing subject of interest for planners and practitioners deals with women's involvement in participatory rural development projects. In the context of processes of privatization and as a result of the state's willingness to disengage in the direct management of rural development, women are asked to reinforce their contribution to the emergence and consolidation of rural organizations, such as development committees, interest groups, self-help organizations, etc. The role of rural women in the process of local development in rural areas is likely to become an important area of concern in the future, both for practitioners and social scientists.

Conclusion

As has been highlighted in this chapter, discourses about rural women in Tunisia have varied according to shifts in development models and policies. In the 1960s and 1970s, gender issues have been rarely taken into consideration. They have been mainly addressed in relationship to social problems in rural areas (lack of education, traditionalism, and population growth), and discourses have focused on the role that women could play in the modernization of the rural family. From the late 1970s to the late 1980s, the re-orientation of national policy objectives towards ensuring food sufficiency and reducing rural/urban inequalities promotes a rising interest in female farm work and an increasing awareness of women's contribution to the household economy in rural areas. As a result of economic restructuring, rural gender studies have particularly flourished during the 1990s. They have been very often fostered by the implementation of rural development projects aiming at increasing women's contribution to rural households incomes, through the promotion of self-employment and female micro-enterprises. Also in relationship to processes of state disengagement and to the implementation of participatory approaches of rural development, women's contribution to the emergence and consolidation of local institutions and rural organizations has become a new area of interest, both for practitioners and researchers. Processes of economic restructuring and shifts in development strategies have thus favoured a growing interest in rural and gender issues and have contributed to a more positive image of rural women, who are no longer viewed as a hindrance but as a potential for development.

References

Abdessalem, K. (1994), *Tableaux statistiques et indicateurs pour la préparation du rapport national sur la femme rurale.* FAO.

Attia, H. (1968), *la femme rurale dans le cadre des réformes de structures.* In Colloque, la participation de la femme rurale au développement économique et social. Tunis: Institut Ali Bach Hamba, du 29 mars au 3 avril, pp. 8-12.

Baffoun-Chouikha, A. (1981), L'accès des tunisiennes au salariat; caractéristiques et incidences. In Michel, A., Fattouma-Diara, A., Agbessi-Dos Santos, H. *Femmes et multinationales.* Paris, ACCT-Karthala, pp. 227-245.

Baffoun-Chouikha, A. (1981-1982), *La recherche en sciences sociales sur la femme maghrébine. Position du problème, tendances, besoins.* UNESCO.

Baraket, M. (1986), La planification familiale en milieu rural: gouvernorat de Mahdia, *Revue Tunisienne des sciences sociales* n° 84/87, pp. 397-421.

Balfet, S. (1982), Travail féminin et communauté villageoise au Maghreb. *Peuples méditerranéens*, n°18, pp. 109-117.

Baccar, H. (1987), *Femmes, développement et environnement,* Communication Présentée au Séminaire: femme, agriculture et développement, Tunis, Union Nationale des Femmes Tunisiennes, 6-7 novembre, 17 p.

Bchir, M. (1973), L'influence sur le taux de la fécondité du statut et du rôle de la femme dans la société tunisienne. RTSS n°32-35, pp. 103-159.

Bchir, B. (1993), Mode de survie et projet de développement de la population forestière de Jendouba: comportement de la femme et perception du changement. *Cahiers du CERES*, Série géographique n° 8, pp. 175-193.

Ben Boubaker, A. (1994), *Perception par les femmes rurales de leur situation et de leur contribution économique: méthodologie et axes de développement.* Ministère de l'Agriculture, Ministère des affaires de la femme et de la famille/CREDIF, FAO, octobre, 47 p. + annexes.

Ben Salem, L. (1993), Introduction au colloque Les transformations actuelles des sociétés rurales au Maghreb. Actes du colloque organisé à Hammamet en Avril. Publications de la faculté des Sciences Humaines de Tunis, Série 7, Volume 5, Tunis.

Ben Sedrine, F. (1968), *Aspects de l'alphabétisation et de l'éducation des femmes en milieu rural.* In Colloque, la participation de la femme rurale au développement économique et social. Tunis, Institut Ali Bach Hamba, du 29 mars au 3 avril, pp. 8–12.

Ben Younes, A. (1968), *L'enseignement de la jeune fille et les écoles professionnelles en milieu rural.* In Colloque: la participation de la femme rurale au développement économique et social. Tunis, Institut Ali Bach Hamba, du 29 mars au 3 avril, pp. 14–18.

Ben Younes, A. (1984), *Rôle des femmes dans l'agriculture et la production vivrière en Tunisie.* Ministère de l'Agriculture.

Ben Younes, A. (1987), *La femme tunisienne et l'agriculture.* Communication au Séminaire Femme, Agriculture et Développement, organisé par l'UNFT, Tunis, 6–7 novembre.

Ben Younes, A. (1987), *Rôle et travail de la femme dans les petites et moyennes exploitations du périmètre irrigué de Jendouba, approche socio-économique.* GTZ.

Ben Younes, A. (1989), *Quelle formation pour la femme rurale?: Une approche participative.* Coopération Tuniso-Allemande, Tunis, 82 p.

Ben Younes, A. et al. (1991), *Formation de la femme rurale.* Ministère de l'Agriculture.

Boukraa, R. (1976), *Notes sur le planning familial et pouvoir au Maghreb.* RTSS n°46, pp. 193–99.

Chabbi, A. (1980), *Population active, revenus et formation dans l'agriculture Tunisienne.* Annuaire de l'Afrique du Nord XIX, CNRS, pp. 89–100.

Chebil, M. (1968), *Développement agricole et évolution de la famille rurale tunisienne.* Institut Ali Bach Hamba, Colloque Maghrébin La participation de la femme rurale au développement économique et social. Tunis.

CREDIF (1994), *Femmes de Tunisie: situation perspectives.* CREDIF.

Delain. B. (1985), *Enquête sur le travail des femmes paysannes: propositions d'intervention dans le cadre du projet de développement rural intégré du Nord-Ouest Tunisien.* Mémoire de fin d'études, Montpellier, CNEARC.

Dimassi. H. (1990), *Etudes et informations sur le travail de la femme: bilan critique* Cahier de l'IREP, n°6, pp. 23–81.

Drira, M. (2000), *Genre et Statistiques: Améliorer la collecte, l'analyse et la présentation des données statistiques et informations différenciées par sexe dans le secteur agricole: aspects conceptuels et méthodologiques; étude de cas en Afrique de l'Ouest et en Afrique du Nord.* FAO.

ERRIF (1981), Femmes et monde rural. *Bulletin d'ASDEAR* n° 3.

FAO, (1995), *Les femmes, l'agriculture et le développement rural.* Rapport National Sectoriel pour la Tunisie. Rome, FAO.

Ferchiou, S. (1978a), Place de la production domestique féminine dans l'économie familiale du sud tunisien. *Revue Tiers Monde,* n°76, oct–dec. pp. 831–844.

Ferchiou, S. (1978b), Répartition sexuelle du travail en milieu tunisien. *Revue Tiers Monde.*

Ferchiou, S. (1978c), Travail des femmes et production familiale en Tunisie. *Nouvelles Questions Féminine,* n°2, fev. pp. 41–55.

Ferchiou, S. (1979), Conserves céréalières et rôle de la femme dans l'économie familiale en Tunisie. In *Les techniques de conservation des grains à long terme.* Paris, éd. du CNRS, pp. 90–97.

Ferchiou, S, (1980), *Conditions de la femme dans les périmètres irriguées de Sidi Bouzid*. Tunis, SIDA.

Ferchiou, S (1981), Women's work and family production in Tunisia. *Feminist Issues*, 1, 2, pp. 55–68.

Ferchiou, S. (1983), L'aide internationale au service du patriarcat. *Nouvelles Questions Féminines*, pp. 47–57.

Ferchiou, S. (1985), *Les femmes dans l'agriculture tunisienne*. Paris, Tunis, Edisud-Cérès Production.

FIDA (1986), *Projet de développement agricole et promotion de la femme rurale dans les gouvernorats du Kef et de Siliana*. Annexes III, composante promotion de la femme rurale. Par K. Rivière. Avril, pp. 59–109.

Gana, A. (1988), *Intégration des femmes au développement agricole en Tunisie*. Banque Mondiale, juin, 40 pp.

Gana, A. and Khaldi, R. (1987), *Le rôle de la femme dans le développement en Tunisie*, communication présentée au séminaire national sur la femme dans l'agriculture, organisé par l'UNFT et le Ministère de l'Agriculture, Tunis, novembre, 9 p.

Gana, A. (1991), Le travail des femmes dans l'agriculture tunisienne. *Annales d'Economie et de Gestion de Tunis* 1, pp. 129–144

Gana, A. (1992), *Le travail des femmes dans les exploitations agricoles familiales: les exemples de Maâmoura/Nabeul, l'Amaeim/Zaghouan et de Mangouche/Bou Salem*. Rapport de recherche, Tunis, juillet, 50 pp.

Gana, A. (1995), *Approche du travail des femmes dans les systèmes familiaux de l'agriculture. Exemple de la Tunisie*. Communication au cours de formation sur le travail des femmes en Méditerranée, nouvelles approche théorique et méthodologique, Programme Med Campus, Tunis du 25 au 29 avril. CREDIF.

Gana, A. (1995), *Travail et temps des femmes dans les exploitations agricoles familiales en Tunisie*. Communication présentée au forum des femmes de la Méditerranée. Le travail des femmes dans les économies nationales et le développement régional. Tunis du 2–4 juin. CREDIF.

Gana, A. (1996), *Femmes rurales de Tunisie*. CREDIF.

Gana, A. (1992), *La promotion du rôle des femmes dans l'agriculture*. Plan d'actions pour les femmes et le développement, Ministère du plan. PNUD.

Gana, A. and Picouet, M. (1998), *Population et environnement en Tunisie*. PNUD/INED.

Gana, A. (1999), *Activités entrepreneuriales des femmes dans le secteur agricole*. CREDIF/ACDI.

Gana, A. (2000), *Micro-crédit et développement participatif*. Analyse de l'expérience menée par le projet de développement rural dans le bassin – versant de Oued Sbaihya (Zaghouan). CES-FAO.

Gallali, F. (1978), Evaluation de l'expérience Antennes Familiales. *Bulletin de l'ASDEAR* n°17, janvier-février, pp. 17–22.

Hotchschild, A. (1967), Le travail des femmes dans une Tunisie en voie de modernization. *Revue tunisienne de Sciences Sociales* 9, pp. 145–166.

Institut Ali Bach Hamba (1968), Colloque Maghrébin : La participation de la femme rurale au développement économique et social. Tunis 29 mars au 3 avril.

Kamoun, A. (1994), *Tableaux statistiques et indicateurs pour la préparation du rapport national sur la femme rurale*. FAO.

Karoui, N. (1980), Etude sociologique sur les ouvrières agricoles dans la région de Mateur. RTSS 63, pp. 92–135.

Karoui, N. (1989a), Rôle et statut des femmes rurales: exemple des îles Kerkennah, RTSS 96/97, pp. 27–66.

Karoui, N. (1989b), Les femmes dans le domaine agricole dans la Tunisie coloniale. RTSS 98/99, pp. 129–151.

Landsberg, G. et al. (1993), *Rapport de consultation: Etude de base sur le problème de l'adéqua-tion formation qualifiante et insertion économique de la fille en milieu rural. Etude de cas des régions de Sidi Bouzid, Sfax, Béja, Zaghouan.* Ministère de la formation professionnelle et de l'emploi, DGFP, DFF, Coopération Tuniso-Allemande. Projet: Formation professionnelle des jeunes filles rurales orientée au marché régional du travail, GTZ. Tunis.

Lamari, M. and Schurings, H. (1999), *Forces féminines et dynamiques rurales en Tunisie.* Paris, L'Harmattan.

Mahfoud-Draoui, D. (1980), Formation et travail des femmes en Tunisie. Promotion ou Aliéna-tion? *Annuaire de l'Afrique du Nord* XIX, pp. 255–288.

Minstere des Affaires Sociales (1985), *L'Insertion professionnelle des jeunes femmes dans les cen-tres du programme du développement rural.* Tunis, Office de la Promotion de L'Emploi et des Travailleurs Tunisiens a L'étranger.

Mahfoud-Draoui, D. (1993), *Paysannes de Marnissa. Le difficile accès à la modernité.* Tunis.

Makhlouf, E. (1994), *Femmes, agriculture et développement rural.* Tunis, Ministère des Affaires de la Femme et de la Famille. CREDIF, Ministère de l'Agriculture, FAO.

Makhlouf, E. (1994), Les activités productives des femmes en milieu rural. In *Actes du sémi-naire: Maghrébines rurales et productrices vers une proposition tunisienne d'accompagnement.* Tunis du 7 au 9 juin. Tunis, IDNS, pp. 41–58.

Ministère de l'Agriculture (1991), *La femme rurale dans les statistiques agricoles.* Tunis.

Ruiz, I. (1994), Du rural à l'urbain. travail féminin et mutations sociales dans une petite ville du Sahel tunisien. *Correspondances, Bulletin de l'institut de recherches sur le Maghreb Con-temporain* 25, pp. 9–14.

Samandi, Z. (1990), La femme dans le processus socio-économique de développement en Tu-nisie. *Cahier de l'IREP, Femme, Famille et Développement* 6, pp. 83–112.

Sahli, S. (1981), Insertion de la jeune fille rurale dans le développement. RTSS 64, pp. 155–162.

Semai, S. (1992), *Promotion de la femme rurale au nord de la Tunisie, expériences du projet Edimo.* GTZ, *Situation de la femme rurale au nord de la Tunisie.* Tunis, Office de l'élevage et des pâturages.

Sanna, A. (1990), Analyse comparative des trois secteurs d'emploi à fort taux de féminisation. In *Evolution de l'emploi féminin dans la région Afrique du Nord et Moyen Orient: cas de la Tunisie.* Tunis, BIRD.

Selmi, A., Khemiri, F. and Sliti, M. (1991), *La formation féminine: une approche participative pour l'intégration de la femme dans les activités du* CPRA. Ministère de l'Agriculture, AVFA, CPRA de Jendouba.

Sethom, H. (1987), Introduction à l'étude des disparités régionales de la fécondité en Tunisie. *Revue tunisienne de Géographie* 15, pp. 159–172.

Tamallah, L. (1982a), *La scolarisation et la formation professionnelle des femmes en Tunisie.* RTSS 66–69, pp. 107–128.

Triki, S. and Makhlouf, E. (1993), *L'activité de la femme rurale et la gestion des ressources natu-relles en Tunisie.* Tunis, Banque Mondiale, CREDIF.

Triki, S. (1994), Les mécanismes d'appui aux activités productives des femmes en milieu rural. In *Actes du séminaire: Maghrébines rurales et productrices vers une proposition tunisienne d'accompagnement.* Tunis du 7 au 9 juin, Tunis, IDNS, pp. 59–83.

Triki, S. (2000), *Le budget-temps des femmes rurales.* CREDIF/PNUD. Tunis.

Union Nationale des Femmes Tunisiennes (1992), *Etude sur l'abandon scolaire des files en milieu rural: rapport final et recommandation.* UNICEF.

Zghal, R. (1986), La formation professionnelle des jeunes filles en milieu rural: principes de base et perspectives d'avenir des jeunes formées. In Psychologie différentielle des sexes. *Cahiers du* CERES, *série Psychologie* 3, pp. 329–343.

Zghal, A. (1965), La modernization de l'agriculture et le statut social de la bédouine. In *Modernization de l'agriculture et population semi-nomade.* CERES.

Zghal, A. (1969), L'élite administrative et la paysannerie. *Revue Tunisienne des Sciences Sociales* 16 pp. 41–52.

Zouari-Bouattour, S. (1998), L'Emploi des femmes. Conférence Nationale sur L'Emploi. Tunis.

Chapter 13

Rural Gender Studies in Spain between 1975 and the Present

Mireia Baylina and Maria Dolors Garcia-Ramon

The year 1975 was a time of important events in the history of Spain, and one of those events was the public appearance of feminism. Women's studies emerged from the feminist movement and was introduced into the universities by academic staff and students in order to make women visible, and to introduce gender into academic disciplines. The universities started offering seminars, and the first research institutes were created, basically specializing in sociology, anthropology and history (Ballarín et al., 1995). This successful start was followed by a more difficult and challenging period between 1978 and 1982, when the feminist movement was of great intensity, but when there was still no political response to feminist demands.

The progressive consolidation of democracy and the modernization of the university system facilitated the establishment of the Women's Institute, a national institution within the Social Affairs' Ministry, in 1983, as well as the approval of the University Reform Law. Both have been very positive for the expansion of women's studies. The Women's Institute has played a crucial role in supporting research and publications. Moreover, the entry of Spain into the European Union, in 1985, prepared the way for the participation of university research groups in European networks. In 1988–90, the government ratified the First Plan for the Equal Opportunities of Women, and from the 1990s onwards, there has been a growth of gender equality institutions in all the regions of Spain.

Women's studies consolidated during the 1990s through the proliferation of study centres, publications, graduate and postgraduate teaching, research projects, doctoral and masters' theses and participation in scientific activities (seminars, conferences). The original 'feminist studies' soon became 'women's studies' and more recently 'gender studies'; a term that is less conflictive to academia, and that has introduced a new theoretical dimension. Gender studies does not focus exclusively on women, but rather on the power relations between men and women. Furthermore, there are now a greater number of disciplines dealing with gender issues, although history, sociology, anthropology, geography, economics and law continue to occupy the forefront (Ortíz et al., 1999).

Gender in Rural Studies

In the context described above, in the second half of the 1980s some universities introduced the gender perspective to rural studies, principally from within geography
and sociology (Garcia-Ramon, 1989; Ballarín et al., 1995). These study programmes
were mostly headed by a small number of professors and were funded by the national
or regional Institutes for Women, or by the Ministries of Agriculture and Education
(Prados, 2000). The reasons for this rather late appearance of rural gender studies in
comparison to Anglo-Saxon or Francophone countries may have been the androcen-
tric character of rural studies in Spain, as well as the more limited and slower development of gender studies in general. In previous decades, when *rural* meant *agrarian*,
women were practically invisible in the activity of this sector. Men were considered
the only actors in the process of agrarian modernization, characterized by the transformation of family farming and its integration into capitalism, and this was reflected
in academic discourse. Nevertheless, some early writings from the 1970s should be
mentioned. García Ferrando (1977) studied the feminization of agriculture and the
adaptive character of rural women; moreover, the position of women in farming received attention through regional and local studies (Zapatero and Jiménez, 1980).

   After these beginnings, there was an intense research activity, reaching its peak in
the 1990s. These studies show, in general, the deep socio-economic changes in rural
areas, focussing on the crisis in family farms and the emergence of the generational
break for rural women, expressed in new activity patterns and new values and life
expectations. In addition, agrarian geography and sociology developed into rural geography and rural sociology, in which all the processes of regional and social change
could be included.

   From a theoretical perspective, we have to underline the use of radical perspectives, which have been very useful in assessing inequalities between men and women
with respect to space and environment. This has facilitated the greater visibility of
women's work within rural households and family farms, a very common issue in rural studies in Spain. In the 1990s, post-modernism, post-colonialism and the so called
'cultural turn' contributed to the inclusion of the theory of difference, which allowed
for the study of the complexity of women's (not woman) life and for the combination of the gender dimension with other causes of difference, such as class, ethnicity,
sexuality or nationality (Garcia-Ramon, 1998). Nevertheless, although certain studies from this epistemological perspective have appeared (Bru, 1995; Garcia-Ramon,
1996), they do not refer specifically to the rural environment (Caballé, 1997).

Research by University Groups

The role of certain research groups has been crucial in the expansion of rural gender
studies. In geography, the research group at the Universidad Complutense de Madrid

is of particular importance. This group examines women's participation in the reactivation of poor rural areas, and, in general, assesses women's role in the rural environment (Sabaté, 1989). In the late 1980s, they began their research on the labour market and rural industrialization: the source of women's labour in the region of Madrid and other Spanish inland regions, such as Toledo (Sabaté, 1992a; 1992b). The consideration of gender as one of the key factors to be taken into account in industrial location was subsequently applied to other regions and economic sectors, and it clarified the relationship between diffused industrialization and the availability of female labour (Martín Caro, 1990; Martín Gil, 1995).

The interest in this subject is shown in the book *Women, space and society. Towards a gender geography* (1995), which clearly advocates a geography that includes the gender dimension in its scientific thought (Sabaté et al., 1995).[1] In this book, issues on gender and rural areas are collected in several chapters devoted to standards of living and women's work in the developed and peripheral countries. More recently, the Madrid group's research has focused on the analysis of the interaction between women and environmentalism (Sabaté, 1999),[2] and specifically on women's contribution to the new ecological agriculture (López, 2000).

Another research group that appeared at the same time is the one at the Universidad Autónoma de Barcelona, which has contributed from its beginnings to the spreading of a gender perspective within geography, initially through the translation of certain basic texts on feminist geography (Garcia-Ramon, 1985) and later through their writings on gender and rural environment and gender theory (Garcia-Ramon, 1989, 1992, 1998).

This research group began its activity with the analysis of women's work and gender relations on family farms in the region of Catalonia, experiencing from the very beginning the difficulties in obtaining significant information on women's work in agriculture. Very soon, the study was extended to other Spanish regions, and the group was enlarged with other researchers from the Universities of Girona, Santiago and Seville (Garcia-Ramon et al., 1994). In 1990 the first Spanish doctoral thesis that relates geography, gender and agriculture was presented in this research context (Cánoves, 1990). Shortly after this, two additional theses followed (Salamaña, 1991; Garcia Bartolomé, 1991), both concerned with women's labour on family farms, and highlighting women's attitude towards the future. The importance of female salaried work in Andalusian agriculture led to the study of day labourers in Andalusia, also analysed within this framework of production and reproduction.

Later, the research group broadened its research agenda, focusing on women and rural employment patterns, a new issue connected with the ongoing processes of economic diversification in rural areas. The group was further enlarged with researchers from the *Universidad de Valencia*, and new issues such as home working (Villarino and Armas, 1997), agribusiness (Domingo and Viruela, 1997; Prados, 1998), rural tourism (Caballé, 1999), and teleworking (Blanco and Cánoves, 1998) have arisen. Recent projects have focused on the comparison of different regions and countries within

Spain and have been the basis for two other doctoral theses (Baylina, 1996; Caballé, 1998). From a methodological perspective, mention should be made of a number of articles examining qualitative methodological approaches within feminist geography (Baylina et al., 1991; Baylina, 1997; Prats, 1998).

In Sociology, the research group at the *Universidad Complutense de Madrid* is particularly important. Its activity is strongly linked to the work requested by the Ministry of Agriculture on the situation of rural women. Studies were carried out within the framework of the First Plan for Equal Opportunities for Women (1988–1990), funded by the Institute for Women. They dealt with aspects such as women's participation in family farms, their role in the labour organization and decision making, their socio-professional future, or their attitudes towards the future of the farm and the rural community (Vicente-Mazariegos, 1991a, 1991b, 1992, 1993, 1994). One of the main contributions to this line of research is a survey conducted among more than 6,000 women in sixteen agrarian regions in order to analyse their activity in family farming. This research group had already published a book in 1991 on the global situation of rural women (Camarero et al., 1991). Other publications about the role of family and gender strategies in the restructuring process within different rural contexts are by García-Bartolomé (1991, 1994) and Sampedro (1996 a,b).

Rural Gender Studies in Scientific Journals and other Publications

The most common outlets for the publication of research results are specialized journals. Among these, *Agricultura y Sociedad* and *Revista de Estudios Agrosociales*, both published by the Ministry of Agriculture, are well-known journals among Spanish social scientists specializing in rural studies.[3] During the last twenty years, both journals have published only a small number of articles (2.1 percent of the total) with a clear gender perspective, and quite a lot of book reviews and articles published outside Spain, especially in Great Britain and the Nordic countries. In general, the articles are by Spanish women (sociologists or geographers) and are mainly empirically oriented or on theoretical contributions developed elsewhere. The paucity of articles may be due to the late arrival of the gender perspective in rural studies within Spain, and the fact that many researchers prefer to publish in feminist journals. In general, these periodicals have been quite sensitive to the topic, although the editorial boards have never contemplated publishing a special issue. In this sense, the limited number of articles is partly compensated by book reviews, which contribute to spreading new ideas and topics.

Thematic issues of certain journals and the contributions of other researchers not linked to research groups have also been very important to the awareness and the diffusion of the gender perspective. In 1988 the journal *El Campo* (an agricultural periodical with a wide circulation) published an issue on 'Woman and agriculture,' with a notably interdisciplinary approach (geography, sociology, psychology, economy, law).

The authors examined many other interesting issues related to rural women, such as property and its unequal distribution, or the financial situation experienced by widows. Many of the articles are empirically oriented, with a limited number being more theoretical. As the journal has a regional character, the issue included case studies from various Spanish regions. This *El Campo* publication was very important as a starting point for research and publications on gender within rural studies. Whilst the journal does not regularly publish articles on gender, it nevertheless published a second special issue on 'Rural Women' in 1995. The topics dealt with in this issue were quite heterogeneous, as was the scale of analysis. However, the articles provided information on a variety of themes of interest (rural women and education and training, the feminization of poverty, women in the fishing industry, rural women in Arab-Muslim societies etc.) in various areas throughout the world. A comparison of the 1988 and 1995 issues shows that the gender perspective had gradually matured and that recent trends in research were broadening the scope of the topic and the approaches made to it.

Another journal, *Documents d'Anàlisi Geogràfica* (DAG) published at the Universidad Autónoma de Barcelona, has also made an important contribution to gender studies in rural areas, although the journal is not, in itself, specialized in rural studies. In 1989, DAG published its first special issue on gender and geography, with special emphasis on 'Agriculture, gender and space.' It included articles by scholars such as Martine Berlan and Janet Momsen, and by pioneers on gender and geography such as Susan Hanson, Janice Monk and Sophie Bowlby. The issue also included the papers presented at the first international workshop on Gender and Rural Areas that took place at the Universidad Autónoma de Barcelona in 1987.

In the 1990s, DAG published two other issues with most of the papers from two international workshops on gender. The first of these (in 1995) included several interesting articles such as those by Sabate (1995) and Oberhauser (1995) on women and rural industrialization in Spain and the United States, and by Townsend (1995) on representing women pioneers in the Mexican rain forest. In the second special issue on 'Women and the environment' (in 1999), there were interesting articles dealing with rural issues such as a critical review of eco-feminism and the Chipko movement (Mawsdsley, 1999), on gender and agro-ecological change in land settlement in Mexico (Townsend, 1999), or on rural development and gender in Western Argentina (Morales, 1999).[4] It is important to point out that this journal publishes articles on gender on a fairly regular basis, and within the context of Spanish geography, it is the geographical journal with the most gender-aware perspective. Unquestionably, this journal has played an important role in the diffusion of the gender approach.

In recent years, other journals have also published special issues on gender and rurality. The journal *Ruralia* (1999) – a multi-disciplinary publication – issued a volume that included articles on topics that had not previously been covered, such as health and the quality of life among rural women, or on questions of power and associational groups in rural areas. *Cuadernos de Geografía*, published at the Universidad de Valencia, published a volume on gender and geography that included several articles on rural areas in Valencia, covering questions of the labour market for women, their identity, and industrial restructuring in rural areas, etc.

During the 1990s, there was a significant rise in the number of publications on gender (in all the social sciences) as well as in the size of the research teams, which published most of their work in this decade. There was also an increase in the number of public institutions and academic centres dealing with the topic. Most of the research and publications in this period are about women and rural development (Garcia-Ramon and Baylina, 2000) in the context of EU programmes such as LEADER and NOW, and with their implementation in specific areas of Spain, although no explicit attention is paid to the impact of such policies on women. Those in charge of the programme implementation sometimes write articles, and the regional or local governments in collaboration with the European Social Fund publish the results. Articles published by academics are usually set within the frame of European socio-structural policies (Pastor and Esparcia, 1998), and references to gender are also rather scarce.

Some publications have been funded by the Institutes for Women in some autonomous communities in Spain, and they deal with the problems of rural women in these regions (Instituto Andaluz de la Mujer, 1994; Institut Català de la Dona, 1994) or with wider topics such as women's contribution to the rural economy (Vera and Rivera, 1999). Moreover, the Trades Union Comisiones Obreras has recently funded several interesting publications, such as the one on women's work in fruit and vegetables processing companies in the Valencia region.[5] Congresses and workshops for different social sciences reflect the growing interest in this approach,[6] and in particular this interest is growing in rural and industrial geography. Congresses include sessions on the topic and, as a result, the proceedings publish several papers on gender (Torres, 1996). Finally, an example of the growing importance of this approach, as well as its integration within the field of rural geography, is the incorporation of the topic into textbooks of rural geography (Garcia-Ramon et al., 1995).

Key Issues and the Evolution of the Research Agenda

Much of the literature is about women's work in farm households. This topic appeared at the time when there was a profound crisis in agriculture and on the family farm, and when the traditional role of women in the patriarchal family was questioned. Women began to develop new strategies that were clearly incompatible with traditional patriarchal values, and early academic writings illustrated the exodus of women from farming as a response to this traditional pattern of gender relations. This is a concern that has appealed to certain specific lines of research over different periods.

Later, several contributions documented and analysed women's work *on* the farm and *for* the farm. These very detailed studies, both quantitative and qualitative, aimed at acknowledging and making visible women's economic and social role. At the same time, also an interest developed in the sexual division of labour within the family farm (labour organization and the decision-making process). These studies underlined the importance of women's input in the farm and confirmed their weak status in the economic domain. Discourse on the invisibility of women's work is very highly developed in rural gender studies in Spain, and covers practically the entire period considered here, reaching its peak, however, in the late 1980s and early 1990s.

Another recurrent topic is the role of women in the framework of new developments in rural areas. Many writings deal with employment alternatives for women in this new context of economic diversification. Their participation in paid work or as entrepreneurs is a common topic of research, and again, the constraints on women's paid employment in the rural labour market are highlighted. Similarly, the study of the reproductive sphere as well as the general conditions of rural women's life are also tackled.

In the last few years there has been a growing interest in the role of women in rural development and in women's agency in a broader sense. Important new subjects have also appeared, such as the relationship between ecological knowledge and women's experience in the environment. In this framework, and within the context of the reform of EU agrarian policy and the Cork Declaration on the future of rural areas, the Spanish Institute for Women organized a national workshop on 'Women, key-actors of rural development.' Many publications have resulted from this, and there has been a significant change: the subject of study has been extended to all women living in rural areas and is no longer restricted to those working in agriculture or living in agricultural households (Prados, 2000). This has meant an enlargement of the gender approach and an acknowledgement of diversity as a key point of analysis.

From a methodological perspective, the whole period considered here is dominated by research with a 'soft' quantitative approach, although qualitative analysis has rapidly increased during the last decade. It can, in fact, be affirmed that the gender approach in rural studies has contributed to the introduction and dissemination of a more qualitative analysis in general rural studies.

In the new century, new issues are emerging and all of them emphasize power relations. Women are playing a key role in the political arena as well as in the dynamics of the production and consumption of agrarian products, the agro-scientific practices of production, the quality of food or the protection of the environment. In this way they can participate as important agents in social movements.

Changes have occurred in rural areas (some of which are highly positive, such as standards of living and working conditions, which now approach the national average), and have transformed social, family and production relations. Moreover, the changes in the ideas about family and household have modified gender roles, gender relations and the sexual division of labour. It is therefore possible that new research will focus on gender identities and how they are transformed within this mosaic of social groups and activities in rural areas.

Spanish rural gender studies is not very strong in its theoretical contributions, but it has produced very valuable empirical studie. Now the gender perspective is becoming more established, a new challenge will arise in the near future: that of combining our rich empirical tradition with new theoretical perspectives. In this way, we will be able to make a significant contribution to the development of international rural gender studies.

Notes

1. This is the first textbook on geography and gender in Spanish, written by Spanish geographers, which meant an important step forward in the production and diffusion of this issue, taking into account that the majority of scientific writing was from the Anglo-Saxon world and therefore reflects that society.
2. Research project *El papel de las mujeres en la agricultura ecológica: producción, transformación y consumo de productos biológicos*, funded by the Institute for Women (1999).
3. *Agricultura y Sociedad* (Agriculture and Society) was founded in 1976 at the end of a long dictatorship. The political context of the moment clearly influenced an editorial policy with a relatively progressive perspective, which was even more explicit after the socialists came to power in 1982. *Revista de Estudios Agrosociales* (Journal of Agrosocial Studies) was established in 1952. In spite of its economic bias, throughout the period we are considering here, some efforts were made to introduce a more social view of rural development, including the gender dimension. In 1999, both journals merged into the new *Revista Española de Estudios Agrosociales y Pesqueros* (Spanish Journal of Agrosocial and Fishery Studies).
4. The authors, who are not Spanish, have been working with the *Universidad Autónoma de Barcelona* team, and their publication and joint work have contributed to the spreading of new topics and approaches.
5. At the end of the 1980s, the same trade union published research on women in the informal economy, with a special focus on rural areas (CCOO, 1987).
6. The last workshop of the Rural Group of the Spanish Association of Geographers (Lleida, September 2000) included, for the first time, an entire session on 'Women and rural restructuring.'

References

Alfonso, C. et al. (1998*), Mujer y trabajo: las empresas de manipulado de frutas y hortalizas en la comunidad valenciana*. Valencia, Confederación Sindical Comisiones.

Ballarín, P. , Gallego, M.T. and Martínez, I. (1995), *Los estudios de las mujeres en las universidades españolas 1975-1991*. Libro Blanco: Madrid, Instituto de la Mujer, Ministerio de Asuntos Sociales, 44.

Baylina, M. (1996), *Trabajo industrial a domicilio, género y contexto regional en la España rural*. Tesis doctoral, Departamento de Geografía, Universidad Autónoma de Barcelona.

Baylina, M. (1997), Metodología cualitativa y estudios de geografía y género. *Documents d´Anàlisi Geogràfica* 30, pp. 123-138.

Baylina, M. et al. (1991), La entrevista en profundidad como método de análisis en geografía rural: mujeres agricultoras y relaciones de género en la costa gallega. *VIColoquio de geografía rural, Asociación de Geógrafos Españoles*, pp. 11–19.

Blanco, A. and Cánoves, G. (1998), El teletrabajo. ¿Alternativa para el mundo rural? *IX Coloquio de Geografía Rural*. Universidad del País Vasco, Asociación de Geógrafos Españoles, pp. 57–62.

Bru, J. (1995), El medi está androcentrat. Qui el desandrocentritzará? Experiencia femenina, coneixement ecològic i canvi cultural. *Documents d´Anàlisi Geogràfica* 26, pp. 271–276.

Caballé, A. (1997), Aproximación al marco teórico y metodológico en la investigación de geografía del género. *Cuadernos Geográficos* 27, pp. 7–27.

Caballé, A. (1998), *Género, agroturismo y contexto regional en España*. Tesis doctoral, Departamento de Geografía, Universidad Autónoma de Barcelona.

Caballé, A. (1999), *L´agroturisme a l´Estat Español. Anàlisi de l´oferta des d´una perspectiva de gènere*. Bellaterra, Universidad Autónoma de Barcelona, Servei de Publicacions, Collecció Materials, 78.

Camarero, L.A., Sampedro, R. and Vicente-Mazariegos, J.L. (1991), *Mujer y ruralidad. El círculo quebrado*. Madrid, Ministerio de Asuntos Sociales/Instituto de la Mujer.

Canoves, G. (1990), *Mujer, trabajo y explotación agraria familiar en Cataluña: un análisis desde la geografía del género*. Tesis doctoral, Universidad Autónoma de Barcelona.

CCOO (1987), *La mujer en la economía sumergida*. Madrid, Confederación Sindical de Comisiones Obreras, Secretaría de la Mujer. Número monográfico sobre Mujeres valencianas: geografía y género, 64, Facultad de Geografía e Historia, Universidad de Valencia.

Documents d´Anàlisi Geogràfica (1989), Número monográfico sobre Geografía y Género, 14. Departamento de Geografía, Universidad Autónoma de Barcelona.

Documents d´Anàlisi Geogràfica (1995), Número monográfico sobre Trabajo, ocupación y vida cotidiana de las mujeres, 26. Departamento de Geografía, Universidad Autónoma de Barcelona y Unidad de Geografía, Universidad de Gerona.

Documents d´Anàlisi Geogràfica (1999), Número monográfico sobre Género y medioambiente, 35. Departamento de Geografía, Universidad Autónoma de Barcelona y Unidad de Geografía, Universidad de Gerona.

Domingo, C. and Viruela, R. (1997), Trabajo femenino en agroindustrias tradicionales. *Cuadernos de Geografía*, Universidad de Valencia, 61, pp. 1–15.

El Campo, Boletín de información agraria (1988), Número monográfico La mujer y la agricultura, 107. Bilbao, Servicio de Estudios del Banco Bilbao-Vizcaya.

El Campo. Boletín de información agraria (1995), Número monográfico La mujer rural, 133. Bilbao, Servicio de Estudios del Banco Bilbao-Vizcaya.

García Bartolomé, J.M. (1991), *La mujer agricultora ante el futuro del mundo rural*. Tesis doctoral, Facultad de Ciencias Políticas y Sociología, Universidad Complutense de Madrid.

García Bartolomé, J.M. (1994), ¿Mujeres agricultoras o mujeres de agricultores? *El Boletín*, 11. Madrid, MAPA (Ministerio de Agricultura, Pesca y Alimentación).

García Ferrando, M. (1977), *Mujer y sociedad rural. Un análisis sociológico sobre trabajo e ideología*. Madrid, Edicusa.

Garcia Ramon, M.D. (1985), El análisis del género y la geografía: reflexiones en torno a un libro reciente, *Documents d´Anàlisi Geogràfica*, 6, pp. 133–143.

Garcia Ramon, M.D. (1989), Para no excluir del estudio a la mitad del género humano: un desafío pendiente en geografía humana. *Boletín de la Asociación de Geógrafos Españoles*, 9, pp. 27–48.

Garcia Ramon, M.D. (1992), Desarrollo y tendencias de la geografía rural (1980-1990). Una perspectiva internacional y una agenda para el futuro. *Agricultura y Sociedad*, 62, pp. 167-194.

Garcia Ramon, M.D. et al. (1996), Orientalisme, colonialisme i génere. El Marroc sensual i fanàtic d'Aurora Bertrana. *Documents d'Anàlisi Geográfica*, 29, pp. 4-34.

Garcia Ramon, M.D. (1998), Gènere, espai i sociedad: una panorámica de la geografía internacional a finals dels anys 90. *Cuadernos de Geografía*, 64, pp. 295-312.

Garcia-Ramon, M.D. et al. (1994), *Mujer y agricultura en España. Género, trabajo y contexto regional*. Barcelona, Oikos-Tau.

Garcia Ramon, M.D., Tulla, A. and Valdovinos, N. (1995), *Geografía rural*. Madrid, Síntesis.

Garcia Ramon, M.D. and Baylina, M. (eds) (2000), *El nuevo papel de las mujeres en el desarrollo rural*. Vilassar de Mar, Barcelona, Oikos-Tau.

Hanson, S. (1992), Geography and feminism: worlds in collision? *Annals of the Association of American Geographers*, 82 (4), pp. 569-586.

Hanson, S. and Pratt, G. (1995), *Gender, Space and Work*. London, Routledge.

Institut Català de la Dona (1994), *La dona en el món rural. Situació i perspectives*. Barcelona, Generalitat de Catalunya.

Instituto Andaluz de la Mujer (1994), *Trabajadoras y trabajos en la Andalucía rural. Situación socio-laboral de la mujer rural en Andalucía*. Sevilla, Instituto Andaluz de la Mujer, Consejería de Asuntos Sociales, Serie Estudios, 3.

López, R. (2000), La agricultura ecológica como una alternativa también para las mujeres, *Actas del X Coloquio de Geografía Rural de España*, Lleida, pp. 490-498.

Martin Gil, F. (1995), *Mercados de trabajo en áreas rurales: un enfoque integrador*. Madrid, MAPA.

Martín-Caro, J.L. (1990), Industrialización rural y condiciones de trabajo de la mujer en el sector textil de Madrid. *III Coloquio de geografía industrial*, Asociación de Geógrafos Españoles, Sevilla, pp. 76-93.

Mawdsley, E. (1999), Repensant chipko: ecofeminisme sota escrutini. *Documents d'anàlisi geográfica*, 35.

Morales, S. (1999), Industria agroalimentaria, desenvolupament rural i gènere a América Latina: treball i vida qüotidiana de les dones assalariades de Santa Rosa(Mendoza, Argentina) *Documents d'anàlisi geográfica*, 35, pp. 121-145.

Oberhauser, A. (1995), Espai, gènere i estrategias econòmiques de la unidad familiar: el treball a domicili de les dones a l'Apalàtxia rural. *Documetns d'anàlisi geográfica*, 26, pp. 147-165.

Ortíz, T. et al. (1999), *Universidad y feminismo en España Ii. Situación de los estudios de las mujeres en los años noventa*. Granada, Universidad de Granada.

Pastor, C. and Esparcia, J. (1998), Alternativas económicas en el ámbito rural interior. El papel de las mujeres en el desarrollo rural. *Cuadernos de geografía*, 64, pp. 527-542.

Prados, M.J. (1998), Trabajadoras de segunda clase. Mujer y empleo en el sector agroalimentario andalúz. *Trabajo*, 4, pp. 171-186.

Prados, M.J. (2000), *Situación socioeconómica de las mujeres rurales en España*. Sevilla, Junta De Andalucía, Consejería de Agricultura y Pesca, monografías 24/00.

Prats, M. (1998), Geografía feminista i metodología: reflexió sobre un procés d'aprenentatge parallel. *Cuadernos de geografía*, 64, Universidad de Valencia, pp. 313-323.

Ruralia. Revista del món rural valencià (1999), Dossier sobre las mujeres en el medio rural, 3. Caudiel-Benáfer, Castellón, Crie.

Sabaté, A. (1989), *Las mujeres en el medio rural*. Madrid, Ministerio de Asuntos Sociales, Instituto de la Mujer.

Sabaté, A. (1992a), La participación de las mujeres en la dinámica social de zonas rurales desfavorecidas. *Desarrollo local y medio ambiente en zonas desfavorecidas.* Madrid, Ministerio de Obras Públicas y Transportes, pp. 123–138.

Sabaté, A. (1992b), Industria rural en Toledo: la incorporación de las mujeres al mercado de trabajo. *Anales de Geografía de la Universidad Complutense,* 12, pp. 277–288.

Sabaté, A. (1995), Mercat de treball femení i industrialització rural a Espanya: relacions amb l'economia global. *Documents d'anàlisi geografica,* 26, pp. 167–178.

Sabate, A., Rodriguez, J. and Diaz, M.A. (1995), *Mujeres, espacio y sociedad. Hacia una geografía del género.* Madrid, Síntesis.

Salamaña, I. (1991), *La dona pagesa, l'oblidada de l'explotació familiar agraria.* Tesis doctoral, Departamento de Geografía, Universidad Autónoma de Barcelona.

Sampedro, R. (1996a), *Género y ruralidad. Las mujeres ante el reto de la desagrarización.* Madrid, Instituto de la Mujer.

Sampedro, R. (1996b), Mujeres del campo: los conflictos de género como elemento de transformación social del mundo rural. In A. García de león (ed.), *El campo y la ciudad. Sociedad rural y cambio social en España.* Madrid, MAPA (Ministerio de Agricultura, Pesca y Alimentación).

Torres, A.M. (1996), El trabajo femenino ante los cambios en los espacios rurales de canarias: el ejemplo del municipio de el tanque. *VIII Coloquio de Geografía Rural,* Universidad de Zaragoza y Asociación de Geógrafos Españoles, pp. 401–412.

Townsend, J. (1995), Es pot parlar en nom dels altres? Es pot, des de fora, representar les dones pioneres de la selva tropical mexicana? *Documents d'anàlisi geogràfica,* 26, pp. 209–218.

Townsend, J. (1999), Gènere i canvi agro-ecològic en l'ocupació de la terra a mèxic. *Documents d'anàlisi geogràfica,* 35.

Vera, A. and Rivera, J. (1999), *Contribución invisible de las mujeres a la economía. El caso específico del mundo rural.* Madrid, Ministerio de Asuntos Sociales, Instituto de la Mujer.

Vicente-Mazariegos, J.I. et al. (1991a), *Situación socioprofesional de la mujer en la agricultura. I Recopilación bibliográfica.* Madrid, MAPA.

Vicente-Mazariegos, J.I. et al. (1991b), *Situación socioprofesional de la mujer en la agricultura. II La mujer en las estadísticas oficiales.* Madrid, MAPA.

Vicente-Mazariegos, J.I et al. (1992), *Situación socioprofesional de la mujer en la agricultura. IV Análisis jurídico.* Madrid, MAPA.

Vicente-Mazariegos, J.I et al. (1993), *Situación socioprofesional de la mujer en la agricultura. V Análisis sociológico.* Madrid, MAPA.

Vicente-Mazariegos, J.I et al. (1994), *Situación socioprofesional de la mujer en la agricultura. III Estudio comparativo con las agricultoras europeas.* Madrid, MAPA.

Villarino, M. and Armas, P. (1997), Globalización y sistemas productivos locales en el textil gallego. *Dinámica litoral-interior. Actas del XV congreso de geógrafos españoles,* Asociación de Geógrafos Españoles, Departamento de Geografía, Universidad de Santiago de Compostela, pp. 983–993.

Zapatero, S. and Jiménez, R. (1980), *La mujer en la explotación familiar agraria: estudio de una zona aragonesa.* Zaragoza, Centro de Investigación y Desarrollo Agrario Del Ebro.

CONCLUSION

Chapter 14

Conclusion: Power, Gender and the Significance of Place

Henri Goverde, Henk de Haan, Mireia Baylina

The chapters in this book have shown that an approach based on the concepts of power and gender opens up new perspectives for understanding the heterogeneous patterns of rural development in Europe. This approach was inspired by the first volume (Halfacree et al., 2002) that resulted from the EU COST project on 'Leadership and Local Power in European Rural Development,' in taking seriously its conclusions and recommendations: '... to what extent are we now able to talk meaningfully of a 'rural Europe' ... or is heterogeneity still the key emergent feature of the rural areas in Europe?' (ibid, p. 273). It is clear that one of the main structural forces transforming the European landscape concerns the construction of a new social contract between agriculture and society (see Cork Declaration, 1996) and, more generally, replacement of the productivist landscape with the emergent landscape of consumption. Research at the local level, as well as theoretical contextualization of the structured circumstances of activities in rural areas have, however, shown that, despite 'a number of general themes and common experiences ... there is still great diversity concerning the issues implicated within rural development' (ibid, p. 274).

Much of the diversity at national level is caused by differences in political configurations, regionalist tendencies and constructions of diverging ruralities in different national contexts. Having come to this conclusion in the previous COST book, we have tried to go a step further by exploring in more detail the different meanings of such crucial concepts as rurality, state and elite. Moreover, we have tried to study the capacity of development policies in the context of concrete changes actually occurring within rural places and around policy issues. This has enabled us to observe how policies can be captured, derailed, or even blocked completely in a local power network constellation, and to show that diversity is not only an issue of differences between polities in nation-states, but also between specific locations. The embeddednes of national rural development policies in local and regional power networks very much determines the outcomes of these processes.

As we have shown in the previous chapters, power is manifested in many ways, and can be based on a variety of resources and structures. We have consciously chosen to focus on gender as one of the important structuring principles in the division and exercise of power in rural areas. As several chapters in this book show, women have long been neglected and marginalized in rural areas. As dependent producers, tied to a domestic mode of production with a strong patriarchal ideology, they have been the silent observer of rural modernization and at the same time a passive subject in the formation of new rural discourses. With the decline of agriculture and the widening space available for women to participate in new economic opportunities and policy programmes, they have become an important agent in the present formation of new ruralities. However, in this process of empowerment they meet important obstacles. The central position of gender is of crucial importance to understanding leadership and local power in rural development, and we are now also beginning to understand the gendered character of spatialities and how women participating in policy discourses challenge them.

This volume has elaborated on power relations in different rural contexts, particularly at the intersection of gender and rurality. As such, all the chapters emphasize in some way or another and in line with the Budapest Declaration on Rural Innovation (2002) that rural innovation oriented to contributing to a new societal contract in rural areas is not neutral in its origins and its outcomes, nor with respect to inequalities of gender and power at local or other levels of social life. Furthermore, it shows that 'place' is still very important in forming contexts and constructing the experiences of social life. It is not surprising therefore that in this volume 'diversity' is also the key word in characterizing rural Europe at the turn of the century. Systematic and comparative research illustrates the cultural diversity of rural Europe and suggests that theories and policies of rural restructuring have to consider these particular social contexts.

The Question of Power

The chapters in Part I allow us to answer some questions concerning the dynamics of power in rural development processes that were raised in the Budapest Declaration on Rural Innovation (2002). How do new forms of social capital and opportunities for political participation affect power relations? The different old and new forms of social capital (family networks, paternalism, new rural partnerships, new local administrative institutions and functionaries) seem to affect local power behaviour in European rural areas in a variety of ways. Some of these forms of social capital explain why the population in rural areas has not been disposed to interventions from a national or international policy-oriented institution. On the one hand, this might suggest a conservative attitude, even inertia, to policies that urge agricultural innovation in order to adapt the local and regional economy to neo-liberal ideology and global market conditions. On the other hand, the refusal to participate in European policy frames can also be interpreted as the result of bottom-up local power networks and conditions in which the firm is not the key unit of analysis, but the family. Then it is the way in which families, often under female leadership, make decisions about continuity and change. In this volume family networks and family strategies have been shown to be the decisive variables in decisions about the continuity of the firm in Spain as well as in (East) Germany. However, even in advanced agricultural settings like the Netherlands and Norway, it has been demonstrated that marginal economic firms often continue because of traditions and as a result of innovations from a gendered origin. In fact, the family often constructs a power configuration that resists a pure business economic approach in farming.

The problem of compliance or resistance, and of continuity and changing power relations, offers a challenge to explain the reproduction of processes of social and systems integration in the rural areas. According to the empirical results described in Part I, an explanation should incorporate several nuances. On the one hand, the dynamics of power reflect ever-changing forms of social capital and political participation. That is why these dynamics of power can contingently cause compliance as well as resistance. In the Basque region, for example, policy-makers often expect to be confronted with resistance to firm-oriented neo-liberal innovations in agriculture. However, the Basque farmers do not think in terms of compliance or resistance. They simply attempt to accommodate their family strategy to the new agro-political conditions (structural power) by mobilizing all the available family resources (episodical power). Furthermore, compliance and resistance can exist at the same time. For example the new local partnerships, bottom-up constructed social capital, in the Finnish context comply with the EU LEADER regime, but resist the processes of sectoralization and specialization initiated by national policy-makers.

On the other hand, the dynamics of power reflect the process that produces and reproduces the circuits of power. The East Germany case study demonstrates that the regime change in 1989 has not resulted in a change in local power networks. On the contrary, the rather hierarchical social structure and the small local elite at the top are maintained as social capital to link the local community to external support systems. In the Norwegian case, however, it seems that the changing macro-social context deeply influences the rural discourses and as such the political agenda at the regional level. However, it seems that discursive formations and deliberations concerning rural and regional policy that started at the national level have constitutive power only if the people concerned are involved and if they recognize and legitimize the new meanings given to their everyday life-world. Policy interventions will be counteracted as long as the policy-makers act from indifference to the people concerned. The Dutch case seems to confirm this Norwegian experience, although in a relatively extreme and contingent way. Although most of the Dutch farmers seem to comply with EU institutions mediated by the Dutch national and regional authorities, animal diseases have produced so many shock events that some farmers and their supporters are now ready to resist these regimes even violently. After the fading away of the neo-corporatist institutions, political authorities usually develop new policies in interactive processes in which all stakeholders participate. Then, public power is no longer dominant and all efforts are oriented to constructing consensus, social and political trust, and social capital. However, during periods of crisis a pure top-down approach by a government setting and maintaining the rules cannot be avoided, even though the stakeholders often resist the legitimacy of such an approach.

Another question in the Budapest Declaration is who controls and who benefits from the perceived change in power relations in rural development? Is the control in the hands of a new or old local elite? Do local and regional organizations in the rural areas have enough trust in the regime of a deliberate democracy and its procedures of interactive policy-making? Concerning these questions, the empirical studies in Part I show some general trends and provide some very pertinent answers as well. For the creation of a new contract between agriculture and society in rural areas, the countries and regions studied generally adopted a multi-actor and multi-purpose, as well as a multi-layered, approach in rural development. As far as policy interventions are relevant, the EU institutions have searched for new (policy) arrangements throughout European rural areas (particularly adaptations in the GAP, introduction of the LEADER programme, and changing criteria in the Structural Funds). The changes in the EU regime often enhance the decentralization of power from the national frameworks of governance to the regional and local level. However, that does not mean that local and regional stakeholders are able to develop new initiatives autonomously. From the perspective of power, local actors are still strongly embedded in national institutions, which often prohibit the construction and the implementation of relatively comprehensive solutions. In most of the case studies, it is claimed that there are two 'real worlds': the world of policy-makers (sectoral bureaucracies) and the world of everyday life. While the policy-makers have a firm belief in the planning and intervention capacities of the state, the local and regional stakeholders, both urban and rural, seem to be involved in reinventing the rural (Goverde and de Haan, 2002, p. 45).

From this perspective the above-mentioned questions concerning the benefits of regime change and the degree of trust in deliberate democracy cannot be answered straightforwardly. Sometimes the answers can only be given after changing the criteria for evaluating the on-going processes (e.g., from 'firm' orientation to 'family' orientation). In other cases, it is clear that local actors benefit (e.g., Finnish new local partnerships after the implementation of subsequent Leader programmes). In another case the regime changed fundamentally (East Germany), but neither the local power structure nor the local elite were transformed. Furthermore, a new local elite, mostly local administrators in combination with representatives of the economic elite, may have become important. The more direct and indirect budget sources are distributed by the regional/rural development system, the more representatives of local governments and entrepreneurs can demonstrate their leadership.

In the case of Hungary (Kovàch, 2002, p. 111), for example, it is well known that regional/rural development money is policy related and that economic elite groups have therefore started to acquire both finance and power in local rural politics. They operate directly and indirectly through their networks and have started to exploit the most relevant political positions, particularly in small and medium-sized settlements where money of the regional development system can only be accessed if the local political authorities are collaborative. This is a specific approach that demonstrates how local agents and local networks manage the challenges produced by structural powers at the national, regional (EU) and global (WTO) level. However, it does not imply inherently a rising productivity through rural innovation.

The result of these multi-actor efforts and multi-level governance is mostly permanent competition for control of power positions and power resources embedded in the structural and systemic layers (as distinguished in the concept of power). However, from the perspective of rural development, it is still an open question whether the multi-functional image of rural areas has been dominated by perceptions of the powerful elites (administrative, economic and scientific agencies), while the market-driven concepts of rurality are perhaps closer to the local people's ideas about the use of space.

Progress in Research on Gender and Rurality (1950–2000)

The reviews of the relevant academic literature on rural gender studies in Part II show an important progress in research on rural gender studies, both theoretically and empirically. This increase is visible in all countries, although the scientific production is higher in the United Kingdom, Norway and the Netherlands than in Spain and Tunisia. Rural gender studies has addressed numerous topical themes and research has been conducted within a variety of paradigms. Still, there are some common strands in understanding the concept of gender as a complex and varying social, cultural, historical and geographic construction, in the acknowledgement of power relations, and the interdependence of space, place and gender.

Initial research was politically motivated and the aim was to focus on rural women's subordination in a context of emerging agricultural crisis and changes in the position of women in society. Scholars attempted to illustrate women's roles in agriculture and their contribution to family households. Thus, research dealt with an assessment of women's socio-economic position and patterns of gender inequality, including the examination of the relationships between paid and unpaid work, the politics of the family, social welfare issues and health matters. Since the outset, critique of the extent and nature of the masculine assumptions underlying the content, theories, methods and purposes of most rural research has been important. Early theoretical writings emphasized patriarchal and capitalist ideologies and this explains why a great deal of the empirical work was devoted to documenting and understanding gender divisions of labour and the ways in which productive and reproductive work are gendered and interdependent. Methodologically, problems have been encountered with the limitations of secondary data sources, such as censuses for their failure to document women's lives on farms, with primary data collection techniques, and with the use of survey methods, which do not offer the subject of research the opportunities to frame their own experiences. In this sense, one of the most important contributions of rural gender studies has been to demonstrate, earlier than social sciences in general, the shortcomings of quantitative methodologies.

Changes in rural areas with the decline of agrarian activities raised new issues on the academic agenda concerning women's roles in rural development, such as off-farm labour markets, women's roles in public space, education and vocational training and the quality of women's lives. In other words, as families became more embedded in the rural economy, trying to adapt themselves to a new endogenous paradigm of rural development, the focus of attention shifted to the involvement of women in this process. The aim to develop the countryside through local empowerment and rural development initiatives gave rise to the use of the concept of women's agency, which again shifted attention from women as victims to their active roles. There was a clear change in focus from women's studies and 'gender roles' to the concept of 'gender relations'. Academic studies in Norway, the United Kingdom and the Netherlands conceptualized rurality and gender as social constructions with the aim of understanding the significance of gender in different situations and contexts. The analysis of gender relations raised questions related to power in different contexts and it became more widely accepted that gender constitutes an important influence on issues such as the distribution and operation of power within rural institutions.

More recently, rural gender studies have become global in scope and strongly influenced by interdisciplinary currents, especially by the growth of 'cultural approaches'. The growing number of studies on rurality as a social construction and its relation to gender has led to the questioning of dominant discourses on rurality and the relation between rurality and sexuality, the discovery of 'otherness' in rural society, and attention for the construction of rural identities and the changing, fluid nature of masculinity and femininity. There is a new focus on the notion of 'gender identity', thus acknowledging the divergent ways in which people experience and perform their gender identities. This implies the relevance of 'pluralities' in gender and rural development. And, of course, in thinking of different social representations of rurality and gender, there is an awareness of the fact that power is bound up discursively in the very social and cultural constructs (Hughes, 1997). Research that explores hegemonic processes, resistance to them and the negotiation of identities is particularly developed in the United Kingdom, Netherlands and Norway. But the proliferation of rural gender studies has created space for the study of many other issues such as poverty and social exclusion, women's participation in ecological production, or the gendered nature of governance and policy.

In summary, the gender perspective has contributed to our knowledge of heterogeneity and diversity in rural society. Also, feminist rural scholarship reflects diversity across cultures and researchers. In this sense, the comparative analysis has shown the 'situated' character of knowledge and has contributed to the understanding of contexts and academic work beyond Anglophone communities.

Explaining Gendered Rural Discourses

The comparative descriptive analysis of the development of gender discourses in rural studies is only a first step. Much more research has to be done, particularly to explain the similarities and differences between gendered relationships in rural development and rural policy throughout Europe. Recent research has given the orientation for such work. For example Roggenband (2002) has developed a research design for comparing similarities and differences in women's movements in different countries such as in Spain and the Netherlands. Her theoretical framework is based on the 'political process approach' (PPA) by Kriesi et al. (1995). This approach has recently become dominant in the field of social movement studies. According to the PPA, four characteristics of the (national) political context determine the chances of facilitation, repression or reform with respect to the rise and fall of social movements:

1. The existing configuration of political cleavages, which determines the emergence of new issues.
2. The formal institutional structure of the state, which determines the degree of openness of a state and its capacity to act.
3. The informal strategy used by the authorities to defend against challengers.
4. The differences in political power between political parties, which determine the possible alliances for social movements.

These four elements of the national political context determine the response of the political authorities to the mobilization of social movements, which in turn determines the movements' strategies, their level of mobilization and the outcomes of the mobilization process. It should be mentioned that the PPA has been used fruitfully in a number of comparative studies. However, it has focused almost exclusively on explaining differences between social movements in countries with stable and established democracies. In general, this approach is too static and cannot account for cross-national similarities.

In order to tackle these methodological problems, Roggenband (2002, pp. 13–21) has pointed to cross-national diffusion[1] between social movements as a possible explanation for international convergences. However, classical diffusion theories are only partly applicable to the field of social movements as they mainly focus on the spread of technological innovations and goods, and hardly deal with more complex processes such as the spread of ideas. Roggenband has proposed that researchers pay more attention to the psychological and symbolic aspects of diffusion. She suggests that diffusion between social movements is a dynamic process of continuing inter-action between so-called pioneers or creators of a paradigmatic frame and the 'fol-lowers' or 'adapters' who use and adapt the frames of pioneers, eventually becoming innovators of new paradigms themselves. She proposes therefore that in social move-ments the object of diffusion is the collective action frame of a pioneer, consisting of a problem-definition or diagnosis and a plan of action or prognosis, which is adapted by following movements.

To test the claims of the PPA and the diffusion thesis, the combined research de-sign has been used to compare the women's movements against sexual violence in Spain and the Netherlands. The results have been relevant for gender studies about and in rurality as well. As Moseley (2000, p. 97) has shown, innovations generally can be characterized by a 'slow-quick-slow' tempo of innovation adoption. All graphs plotting the pace of adoption of specific innovations such as hybrid seed corn by American farmers in the 1930s, or the Minitel information service by French house-holds in the 1980s, show similar 'S'-shaped curves. However, scholars who have adopted the claims of PPA have wrongly neglected to note that the women's move-ment does not fit their definition of instrumental and political movements (see also Bock, 2002). They have also neglected to see that women's movements aim to discuss what is considered private and public, and therefore what is political and should be the subject of state intervention. Furthermore, the diffusion thesis assumes that the problem to be studied can be defined as universal. Then the local conditions become relevant in order to investigate whether the universal phenomenon has been diffused differently in these contingent contexts (van Hoven-Iganski, 2000; p 192). This brings the fundamental discussion about power and gender back on the agenda.

Although this book has produced a lot of interesting material for a more global view of rurality in Europe in the face of power and gender, from a methodological perspective it can be concluded that more research about the conceptual relationship between power and gender is necessary in order to construct a common basis for deeper comparative research. Whereas the concept of power has been elaborated and assessed following the classical 'three dimensions of power' (with reference to Dahl, Bachrach and Baratz, and Lukes) and a Foucauldian approach of discourse analysis, the relationship between power and gender is still underexposed. Of course, the de-velopment of rural gender studies has shown the essence of power, domination and resistance in male and female relationships. Still, we agree with Allen (1999, p. 129) that after a reflection on the concept of power[1], a gender orientation to power need not only pay attention to 'power over' and 'power to,' but also to 'power with,' if it is the aim of gender studies both to explain and to describe diverse male and female inter-ests concerning dominance, resistance and solidarity.

Notes

1. 'Diffusion' refers to the spread of something within a social system. The key word here is 'spread,' and it should be taken viscerally (as far as one's constructionism permits) to denote flow or movement from a source to adopter, paradigmatically via communication and influence. We use the term 'practice' to denote the diffusing item, which might be a behaviour, strategy, belief, technology, or structure (Strang and Soule, 1998, p. 266, cited in Roggenband, 2002, p.13).
2. Amy Allen (1999, p. 119–135) concludes that the following authors should be contested concerning their approach to power: Foucault (insufficient attention to normative distinctions), Butler (too narrow a focus on the discursive as the dimension through which power is exercised and reproduced), Arendt (power as acting by people in concert is too communicative).

References

Allen, A. (1999), *The Power of Feminist Theory. Domination, Resistance, Solidarity*. Boulder, Colorado, Westview Press.

Bock, B.B. (2002), *Tegelijkertijd en tussendoor. Gender, plattelandsontwikkeling en interactief beleid* (Ph.D. thesis, Wageningen University).

Halfacree, K., Kovách, I and Woodward, R (eds) (2002), *Leadership and Local Power in European Rural Development*. Aldershot, Ashgate.

Hoggart, K., Buller, H. and Black, R. (1995), *Rural Europe: Identity and Change*. London, Edward Arnold.

Hoven-Iganski, B. van (2000), *Made in GDR. The Changing Geographies of Woman in the Post-Socialist Rural Society in Mecklenburg-Pommerania*. Utrecht/Groningen, Netherlands Geographical Studies 267.

Hughes, A. (1997), Rurality and 'culture of womanhood'; domestic identities and moral order in village life.' In: Cloke, P. and J. Little (eds) (1997), *Contested Countryside: Cultures, Otherness, Marginalisation and Rurality*. London/New York, Routledge, pp. 123–137.

Kriesi, H.P. et al. (1995), *The Politics of New Social Movements in Western Europe. A Comparative Analysis*. Minneapolis, University of Minnesota Press.

Moseley, M.J. (2000), Innovation and rural development: some lessons from Britain and Western Europe. *Planning Practice & Research* 15 (1–2), pp. 95–115.

Roggenband, C. (2002), *Over de grenzen van de politiek. Een vergelijkende studie naar de opkomst en ontwikkeling van de vrouwenbeweging tegen seksueel geweld in Nederland en Spanje*. (Ph.D thesis, KU Nijmegen), Assen Van Gorcum.

Index

Printed and bound by CPI Group (UK) Ltd, Croydon, CR0 4YY

22/10/2024

01777640-0002